UNBELIEVABLE

UNBELIEVABLE

7 Myths
About the History and Future
of Science and Religion

MICHAEL NEWTON KEAS

I S I
BOOKS
Wilmington, DE

Cataloging-in-Publication Data is on file with the Library of Congress.
ISBN: 978-1-61017-153-3

Published in the United States by

ISI Books
Intercollegiate Studies Institute
3901 Centerville Road
Wilmington, Delaware 19807-1938
www.isibooks.org

Manufactured in the United States of America

For my extended family, especially my
grandfather Newton Watson Keas

Also for John, Edna, Chris, Martin, and Loetta

Contents

Introduction

THE TRUTH IS OUT THERE

As much as any other human enterprise, science needs stories—
to portray its origin and development, to glorify its heroes, to
dramatize its methods and values, and to proclaim its place in
human civilization.
> —*Gregory Schrempp*, The Ancient Mythology of Modern
> Science: A Mythologist Looks (Seriously) at Popular
> Science Writing *(2012)*

As members of the human race, we are currently the only highly
intelligent beings in our corner of space. We have looked around,
peered out from our earthly platform, and discovered the universe. We have mapped it and deciphered much of its history.
Maybe we are the first to do so. Maybe the universe...has slept
in ignorance of itself these aeons until suddenly in a few years
mankind's curiosity...has opened its cosmic eyes and the universe has seen itself for the first time. Maybe we are the brains
of this outfit.... We are like Robinson Crusoe, stranded on a
cosmic island, not knowing whether or not we are alone until we
can see the footprints in the sand.
> —*Paul Hodge*, Concepts of Contemporary Astronomy *(1979)*

Scientists enjoy telling stories. They tell stories about, among
other things, the quest to understand the universe—stories that

sometimes have implications for belief or disbelief in God or a spiritual heaven.

Too often, however, these stories are false. They are nothing but myths. And yet some leading scientists and science writers offer these stories as unassailable truth. The myths make their way into science textbooks—which is a useful measure of a myth's influence, as we will see in this book. They also enter popular culture, whereby the myths pass as accepted wisdom.

This book examines seven of the most widespread and influential science myths. These myths concern the relationship between science and religion, especially Christianity. More specifically, the stories depict religion and science as being perpetually at odds.

In our time, history is commonly abused for polemical purposes, wielded as a long sword in the culture wars. The cultural wave known as the New Atheism has grossly distorted the history of science and religion. Yet atheist activists are far from alone in misleading the public. Prominent scientists and science educators—including influential popularizers such as Bill Nye the "Science Guy," the late Carl Sagan, and Neil deGrasse Tyson—lead the chorus of voices portraying Christianity as anti-science, and science as liberating us from religious dogma.

But that narrative is deeply flawed. Sometimes those who champion the misleading narrative acknowledge its holes. As chapter 9 will show, one prominent historian and philosopher of science thinks that it might be permissible to offer an inaccurate history of science for an ideological-political "greater truth." He even suggests granting science educators an "artistic license to lie."

Taking such a morally ambiguous approach to the history of science is disturbing and dangerous. This book corrects the record, revealing the truth that the proliferation of myths has obscured.

So in addition to debunking seven pervasive myths about science and religion, I will present a more accurate, and I hope more interesting, history of how science and religion interact.

The popular science fiction TV series, and later movie, *The X-Files* had a famous tagline that suggested a conspiracy to cover up evidence of extraterrestrial life and the paranormal: "The Truth Is Out There."

It certainly is—but not in the way the tagline suggests. The truth about science, and even about the likelihood of extraterrestrial life, is there for us to see. But first we need to clear away the myths that have done so much to confuse our understanding about science, religion, and our place in the universe.

Seven Myths, Debunked

You've probably heard many of the myths I examine in this book. These seven stories are widely known and accepted even though they rely on exaggerations at best and completely distort the facts at worst.

The first six myths relate to the history of science and religion. The seventh is subtly connected to the six historical myths but imagines a future far removed from what we know today.

Myth #1: Premodern scholars in the Western tradition thought the universe was small—a cozy little place just for human benefit. Modern science displaced this Church-sanctioned belief with a vast cosmos that revealed humans to be insignificant.

The truth: Even ancient thinkers recognized that the earth was tiny in relation to the immense cosmos. In any case, size doesn't necessarily mean significance, as many theologians and philosophers recognized.

Myth #2: The medieval Catholic Church suppressed the growth of science, causing Europe to descend into the "Dark Ages."

The truth: The medieval Catholic Church positively influenced science and other intellectual pursuits. There *were* no "Dark Ages."

Myth #3: Because of Church-induced ignorance, European intellectuals believed in a flat earth until Columbus proved earth's roundness in 1492.

The truth: Ancient and medieval intellectuals in the Western tradition had many evidence-based reasons for belief in earth's roundness.

Myth #4: Giordano Bruno became a martyr for science when the Catholic Church burned him at the stake because he supported

Nicolaus Copernicus's contention that the sun, not the earth, occupied the center of the universe, and because he believed in extraterrestrial life.

The truth: Bruno's execution occurred almost entirely for theological reasons, not scientific ones.

Myth #5: The Church jailed Galileo Galilei because it rejected his telescopic observations and rational arguments that had proved the Copernican system.

The truth: Most early modern astronomers up through the mid-seventeenth century resisted a moving earth primarily for scientific, not theological, reasons. Galileo failed to prove that earth orbited the sun (that came later).

Myth #6: Copernicus demoted humans from the privileged "center of the universe" and thereby challenged religious doctrines about human importance.

The truth: Copernicus (a canon in the Catholic Church) and most of his scientific successors up through the nineteenth century considered his sun-centered astronomy to be compatible with Christianity and human exceptionalism. In fact, early Copernicans viewed earth's new location not as a demotion for humanity but rather as a promotion, out of the *bottom* of the universe.

Myth #7: If and when we encounter extraterrestrial life, it will deal the death blow to certain religions, especially Christianity, with its doctrine of the unique incarnation and redemptive work of God's Son on earth. Any ET capable of traveling a vast distance to earth would have superintelligence, technology indistinguishable from magic, and moral-spiritual insights that would trigger global religious reorientation.

The truth: Many Christian thinkers have been open to the possibility of extraterrestrial life, and neither a single pope nor a major church council ever declared these ideas heretical. ET and Christianity are potentially compatible.

These myths did not arise coincidentally. In part two of this book we will dig deeper, exposing the sources, real and imagined, for the origin and perpetuation of the myths. Science fiction has contributed, as have popular science communicators. In the process, we will see that religion and science need not war with each other—and didn't for centuries. We will also see that these misleading narratives did not gain traction until relatively recently.

The *Other* Kind of Myth

So far I have discussed the word *myth* in its common meaning—as a false story. Another meaning of the term will concern us greatly.

The Harvard scholar Michael Witzel defines a myth in this second sense by noting that it "contains and brings out such images of the world (a cosmology), of past and present society (a history and sociology), and of the human condition (an anthropology) as are eminently constitutive of the . . . society in which that narrative circulates, or at least where it circulated originally," and that "may be invoked (etiologically) to explain and justify present-day conditions."[1]

Put more simply, myths are imaginative archetypal stories that shape a culture's identity and dominant worldview.

So when, in the first epigraph to this chapter, the anthropologist Gregory Schrempp says that, "as much as any other human enterprise, science needs stories," he is talking about myths in this other sense of the term.

And astronomer Paul Hodge's soaring narrative in the second epigraph offers a textbook example of how such a myth operates in the culture of science. Hodge's story is not a myth about the god Zeus but rather one about humanity. The hero of the story is intelligent life, especially smart scientists who study the origin and structure of the cosmos. Thanks to these scientists, "the universe has seen itself for the first time."

A myth in this second, scholarly sense of the term plays the role of imaginative, archetypal story whether it is true or false. The last myth we investigate in this book, about the prospects for and implications of discovering superintelligent life, relies on flawed assumptions and ignores some of the laws of nature (which technology can never overcome). But it also functions as a myth in the scholarly sense. It fills the craving for human significance after many in the world of science have tried to dismiss religious belief. If Copernicus demoted humans, the myth of extraterrestrial enlightenment now fills that void for many people.

Scientists, like all humans, love stories. Most of us experience the world as if our lives are meaningfully going somewhere, tracing the arc of a story. We discover and create. We tell about what we have discovered and created. These are our stories. The big stories, the ones about our place in the cosmos—how we got here and why—are our great myths.

Part 1

7 MYTHS ABOUT SCIENCE AND RELIGION AT WAR

1

BIGGER IS BETTER

To the ancient observer the earth seemed vast and immobile. . . . It was natural, of course, for early man to attach such central importance to his earth. His experience had not prepared him to accept the sky as virtually empty space extending to such an unimaginable—perhaps infinite—depth that the earth in comparison becomes a mere point in the cosmos.
  —*George Abell*, Exploration of the Universe *(1969), 2*

The earth has, to the senses, the ratio of a point to the distance of the sphere of the so-called fixed stars.
  —*Ptolemy*, Almagest, *Book I (ca. AD 150)*

One of the earliest myths about science and religion to appear in English-language science textbooks would have us believe that premodern scholars in the Western tradition thought that the universe was small—a cozy little place *just* for human benefit—and that when science showed otherwise, it dealt a blow to faith.

This myth has filtered down to pop culture. In the 1997 movie *Contact*, based on the novel by astronomer Carl Sagan (1934–1996), Dr. Ellie Arroway (Jodie Foster) looks to the heavens and estimates, given cosmic immensity, that there are "millions of civilizations out there." This was Sagan's own view. Her liberal theologian quasi-boyfriend, Palmer Joss (Matthew McConaughey), responds, echoing

Ellie's father earlier in the movie, "Well, if there wasn't, it would be an awful waste of space." Ellie agrees with an "amen." Near the movie's end, in response to a child's question about other persons "out there," Ellie repeats the intuition about all the wasted space if there's no extraterrestrial life (ET).

Later, ET will get its fair share of space in our story. For now, let us focus on space itself, and how much stuff is inside it. What does this mean for the value of humanity? We know the universe is big—really big. Did people in the Western tradition have to wait till *modern* science to know this, while shedding historic Judeo-Christian views en route? Actually, no.

Despite what the astronomer George Abell (1927–1983) asserted, ancient sky watchers didn't believe that the earth was very large compared to the total volume of the universe. Consider that Ptolemy, the most sophisticated and influential ancient astronomer, taught that earth was merely "a point," virtually dimensionless, compared to the vast distance to the stars.

But Abell and his followers have carried the day.

Abell and Bill Nye Say Tiny Humanity Sucks

Like George Abell, celebrity TV science educator Bill Nye, the "Science Guy," likes to say that modern astronomy has revealed the insignificance of humanity. Nye is CEO of the Planetary Society—the world's largest nonprofit space interest group, cofounded by Carl Sagan, who was the premier science popularizer before Nye. In the last minutes of his 2010 "Humanist of the Year" acceptance speech, Nye—speaking for science and all humanity—delighted the American Humanist Association with a trademark rhetorical flourish: "I'm insignificant.... I am just another speck of sand. And the earth really in the cosmic scheme of things is another speck. And the sun an unremarkable star.... And the galaxy is a speck. I'm a speck on a speck orbiting a speck among other specks among still other specks in the middle of specklessness. I suck."[1]

Nye's audience laughed approvingly, no doubt because they believed that "I suck" really means "religion sucks."

Nye, as a famed science communicator, has clearly crafted an anti-theistic message and persona. Here, too, he shares similarities with George Abell. In a 1977 interview, Abell proudly reported that his agnostic father had founded the Hollywood Humanist Society. "I was brought up heathen," he declared. Did this affect his work as an astronomer? Indeed, for he admitted a bias against the idea of the universe's origin in a big bang, or cosmic singularity, with its implications favorable to theism: "Philosophically I was attracted to the steady-state theory, because it's philosophically easier to envision an infinite universe, infinitely old, than one with such a singularity. On the other hand, I have to admit that the observations don't look very favorable to a steady-state universe."[2] The steady-state theory, which denied a cosmic beginning, was widely regarded as falsified by the time of Abell's 1977 interview. Virtually nobody defends it today.

Later in the interview Abell expressed his anti-theistic view more candidly: "I think most people want very much to have something concrete to hang their hats on and believe in, which is why people are religious. Where did the universe come from? Well, it's much easier to say, 'God made it,' than it is to say, 'I don't know.' But that bothers me a whole lot more than even a singularity; I mean, where did God come from?"

Abell, the agnostic son of an agnostic father, failed to understand the Judeo-Christian view of God, namely that if such a God exists, then he is uncaused, without beginning or end. When an anti-theist like Abell attacks Abrahamic religion with a snarky objection—"Where did God come from?"—he reveals a fundamental misunderstanding about what Christians and Jews mean by "God." Furthermore, advocates of naturalism like Abell inevitably, if only implicitly, believe in some other entity or set of entities that play the god-substituting role of the uncaused self-existent source of all things—be it the material world, a multiverse-generating mechanism, or whatever.

In reference to himself and most of his UCLA faculty friends, Abell confessed, "I think we're all agnostic or atheistic or whatever you want to call it." Was his social world *that* small?

Abell admitted that his anti-theistic bent affected his evaluation of the evidence for a cosmic beginning. It is also plausible that this perspective played a role in his perpetuation of the myth of premodern cosmic smallness, which suggested that science progressively liberated us from religious dogma. In his book *Exploration of the*

Universe, Abell asserted (inaccurately, as we will see in chapter 2) that in "medieval Europe no new astronomical investigations of importance were made," and that "rather than turning to scientific inquiry the medieval mind rested in acceptance of authority and absolute dogma."[3] Christianity retards the growth of science? Ironically, the agnostic Abell was also on the wrong side of history when it came to twentieth-century cosmology: the evidence for the big bang cosmic beginning defeated the steady-state beginningless universe. He missed the big deal.

Chronological and Spatial Snobbery Exposed

Few scholars have better understood "the medieval mind" than (the once-agnostic) C. S. Lewis, a contemporary of George Abell's agnostic father. Lewis made his professional reputation not as a Christian apologist but as a scholar of early English literature. His classic introduction to medieval and Renaissance literature, *The Discarded Image*, was based on a deep familiarity with the original sources. In the book, Lewis summarized the medieval cosmic scale: "Earth was, by cosmic standards, a point—it had no appreciable magnitude. The stars...were larger than it."[4] He cited numerous medieval authors to document this assessment. For example, the popular medieval work *South English Legendary* stated that if a man could travel upward at a rate of about forty miles a day, he still would not have reached the *stellatum* ("the highest heven that ye alday seeth") in eight thousand years. The fact that this medieval stellar distance is much smaller than modern estimates gives the "premodern small cosmos" myth the superficial appearance of truth. But, as Lewis deftly argued: "For thought and imagination, ten million miles and a thousand million are much the same. Both can be conceived (that is, we can do sums with both) and neither can be imagined; and the more imagination we have the better we shall know this."[5]

The crucial difference, Lewis explained, is that the

medieval universe, while, unimaginably large, was also unambiguously finite. And one unexpected result of this is to make the smallness of earth more vividly felt. In our universe she is small, no doubt; but so are the galaxies, so is everything—and so what? But in theirs there was an absolute standard of comparison. The furthest sphere...is, quite simply and finally, the largest object in existence. The word "small" as applied to earth thus takes on a far more absolute significance.[6]

Although the modern celestial distance numbers are bigger than the medieval, the bottom line is similar: the unimaginable largeness of the cosmos (and smallness of earth) in the minds of both medieval and modern people. Further, Lewis showed that earth's smallness was more vividly perceived within the medieval framework than it is within modernity.

Copernicus Made Our Big Universe Bigger

What did the founders of modern science judge to be the human implications of the newly expanded Copernican cosmic scale? To appreciate the answer to this question, we must visualize an astronomical puzzle that Copernicus faced.

Nicolaus Copernicus (1473–1543) was the first to develop a complete mathematical astronomical account of how the earth, along with other planets, revolved around a stationary sun. Considering earth a planet created a new problem for astronomers. As earth travels around the sun, stars A and B in Figure 1.1 (*see next page*) should appear at different angular distances from each other depending on where earth is located in its annual motion. This expected observation is called stellar parallax. Figure 1.1 shows earth when it is close to stars A and B, producing a 30° separation. Four months later, earth's more distant location in relation to these two stars would make them appear only 15° apart. But prior to 1838, astronomers failed to detect any such stellar parallax. Before Newton incorporated Copernican astronomy into his physics in 1687, this failure was widely regarded as evidence against the heliocentric theory.

To explain away the observed *lack* of expected stellar parallax in a heliocentric world system, Copernicus revised the Ptolemaic cosmological scale of distances by asserting that the diameter of earth's *annual motion* (rather than just the diameter of earth itself, as Ptolemy

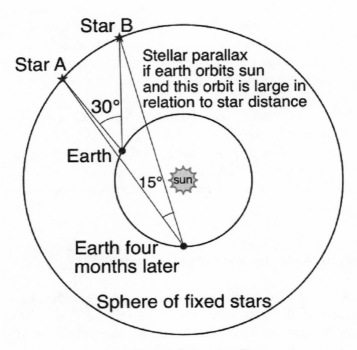

Figure 1.1: Stellar parallax

had argued) is like a point compared to the distance to the stars (see Figure 1.2). This Copernican response to the evidential challenge also gave the planetary part of the universe a much smaller space to occupy compared to the size of the whole cosmos, which early modern astronomers thought was bounded by the sphere of fixed stars. What did these astronomers perceive to be the human implications of this revised cosmic scale?

The astronomer Johannes Kepler (1571–1630) emphatically repudiated the suggestion that the Copernican doctrine of a huge universe (compared to the size of earth's orbit) challenged theistic notions of human significance. In a 1598 letter, he acknowledged that "we are puny" compared to the size of the cosmos, but "one must not infer from bigness to special importance." He explained:

> If the planets were the most unimportant part of the world because the entire planetary system practically disappears when compared with the fixed star system, according to this same

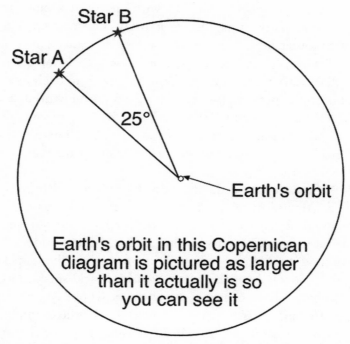

Figure 1.2: The Copernican response to the observed lack of stellar parallax

argument man would belong to the absolute trifles in the world, since he can in no way be compared with the earth and this, the earth, in turn cannot be compared with the world of Saturn. Yes, the crocodile or the elephant would be closer to God's heart than man, because these animals surpass the human being in size.[7]

The Copernican inflation of cosmic real estate between the orb of the last known planet (Saturn) and the stars did not undermine human importance, Kepler argued. Size and significance do not necessarily correlate.

Spaceships of the Imagination

The renowned French mathematician, scientist, and philosopher Blaise Pascal (1623–1662), in his unfinished posthumous defense of Christianity known as *Pensées* (Thoughts), invited us to recognize earth "as a single speck compared with" its orbit around the sun. Going further in his imaginative journey, he declared it a "marvel that this vast orbit is itself only a very delicate point compared to" the distance to the stars.[8] In this, Pascal took for granted the space-inflationary revision Copernicus had made to explain away the absence of stellar parallax (see Figure 1.2). A journey through Copernican immensity, thought Pascal, helps us humbly to worship God. (Scholars who think Pascal did not accept the Copernican system should read my Appendix B, where I advance a new argument for Pascal's Copernicanism.)

"But," Pascal continued, "if vision stops there," with the spatially expanded Copernican view of the universe (which was bigger than the Ptolemaic one), we would not fully appreciate our smallness. Rather, "let imagination pass beyond... this visible world" in order to conceive how it is "but an imperceptible feature in the ample bosom of nature." The whole of the natural world, Pascal speculated, "is an infinite sphere whose centre is everywhere, whose circumference is nowhere. In short, it is the most sensible sign of the omnipotence of God—let our imagination lose itself in this thought."[9] Human significance, Pascal suggested, is found in relation to God, *not* by reference to any particular spatial magnitude or cosmic location, whether seemingly central or peripheral.

Pascal's most famous saying about the great magnitude of space is often cited as if it licenses a godless universe: "The eternal silence of these infinite spaces frightens me."[10] But this is an isolated fragment in the original manuscript and so its immediate context is unknown. Given the total context of Pascal's *Pensées*, this enigmatic utterance more likely was intended to convey the ubiquitous human emotional reaction (including Pascal's own) to cosmic immensity. That does not negate the strong case for God found throughout his unfinished manuscript. Indeed, early modern Christian thinkers such as Pascal embraced the expanded Copernican cosmos in a faith-affirming manner. They recognized the greatness of God reflected in the huge cosmos—and the utter smallness of humanity—as a prompt to humble ourselves in worship of God.

Bill Nye's "speck orbiting a speck among other specks among still other specks in the middle of specklessness," while entertaining to a twenty-first-century audience, is trite in comparison to Pas-

cal's expansive vision of the natural world. Although Pascal lacked
compelling reasons to conclude that the universe is actually infinite
spatially, his mind found the idea compatible with the Son of God's
coming to earth to redeem those who would accept his sacrifice for
sin. Jesus was born in a tiny manger located in an obscure village
within a geopolitically minor culture. The all-encompassing God
became a humble speck, Pascal believed. The cosmological scandal of
undeserved divine favor![11]

Speculation about specks in lots of specklessness has a longer
imaginative history than Bill Nye might realize. Our next Christian
speculator is the Dutch physicist and astronomer Christiaan Huygens
(1629–1695), another remarkable star of early modern science. In his
posthumous *Cosmotheoros* (1698), he wrote that earth, when seen in
Copernican astronomical context, is but a "small Speck of Dirt." He
believed—based on a weak analogical argument—that the vast cosmos was filled with countless inhabited planets. Huygens wrote:

> What a wonderful and amazing Scheme have we here of the mag-
> nificent Vastness of the Universe! So many Suns, so many earths,
> and every one of them stock'd with so many Herbs, Trees and
> Animals, and adorn'd with so many Seas and Mountains! And
> how must our Wonder and Admiration be encreased when we
> consider the prodigious Distance and Multitude of the Stars?[12]

He suggested that this cosmic perspective has a few moral and
theological implications:

> We shall be less apt to admire what this World calls Great,
> shall nobly despise those Trifles the generality of Men set their
> Affections on, when we know that there are a multitude of such
> earths inhabited and adorned as Well as our own. And we shall
> worship and reverence that God the Maker of all these things;
> we shall admire and adore his Providence and wonderful Wis-
> dom which is displayed and manifested all over the Universe.[13]

Humility before the creator, not godless dethronement of humanity, is the appropriate response to a huge cosmos filled with varied
forms of life (including intelligent ET), Huygens argued.[14]

Similarly, William Whiston, the Lucasian Professor of Mathematics at Cambridge University in the wake of Isaac Newton's 1701 resignation, took in the immensity of the universe seen through the telescope as an expression of "the inexhaustible Power of Almighty God."[15]

In seventeenth- and eighteenth-century English astronomy literature, I find no indication that the increasingly larger estimates of cosmic dimensions became grounds for debilitating doubts about human significance. In fact, one of the most quoted biblical passages in eighteenth-century astronomy literature is Psalm 8:3–5:

When I look at your heavens, the work of your fingers,
the moon and the stars, which you have set in place,
what is man that you are mindful of him,
and the son of man that you care for him?
Yet you have made him a little lower than the heavenly beings
and crowned him with glory and honor.

King David sang of how insignificant we feel as we gaze upon God's vast cosmos. Why should God care about puny humans? Yet God granted us slightly less than angelic celestial status. This song provokes great gratitude for God's honor bestowed on humanity. Similarly, in Psalm 103:10–12, David invokes huge celestial distances to illustrate the magnitude of divine forgiving love toward God's authentic followers.

He does not deal with us according to our sins,
nor repay us according to our iniquities.
For as high as the heavens are above the earth,
so great is his steadfast love toward those who fear him.

David is not the only Old Testament writer to use cosmic immensity to illustrate the majesty of God. For example, Isaiah reports the following divine utterance: "As the heavens are higher than the earth, so are my ways higher than your ways and my thoughts than your thoughts" (Isaiah 55:9). The vast majority of Copernican astronomers embraced these biblical perspectives, until a notable falling away over the past two centuries.

The Big Myth Arises

A rare early version of the "premodern belief in a small cosmos" myth in English-language astronomy textbooks appeared near the end of the eighteenth century. George Adams (1750–1795), who was King George III's mathematical instrument maker, wrote in his *Astronomical and Geographical Essays* that "no sooner was an idea formed of the vast extent and greatness of the universe, with respect to this earth, than mankind began to conceive it would be more rational that the earth should move, than the whole fabric of the heavens."[16] Adams assumed that educated people in ancient and medieval times generally believed in a small cosmos. But he got the story backwards: early modern scientists did not conceive of the heliocentric model to account for increased estimates of cosmic dimensions; Copernicus increased his calculations of cosmic size as an ad hoc response to criticisms that his heliocentric theory could not account for the lack of stellar parallax.

In any case, note that Adams did not frame his argument as a point scored against religion. This is unsurprising given that he had a reputation for harmoniously combining faith and science so as to "combat the growing errors of materialism, infidelity, and anarchy."[17]

The first mature instance of the "premodern belief in a small cosmos" myth that I have identified in astronomy textbooks occurs in George G. Carey's *Astronomy* (1824), based on lectures that stretched back to 1815. Carey, a little-known British science teacher, proclaimed that "the most general opinion" in medieval times included the idea that the "heavenly bodies" were "at no great distance from the earth; and that all the celestial bodies were created solely for its use and ornament."[18] The medieval Catholics who allegedly held this belief were rebuked by the progress of science. Carey, who professed allegiance to Christianity, probably did not realize that many later writers would turn his anti-Catholic propaganda into a case against Christianity as a whole.

Two instances of the cosmic-size myth from a century later are worth noting: Simon Newcomb's *Astronomy for Everybody* (1902) and Camille Flammarion's *Astronomy for Amateurs* (1904). Newcomb (1835–1909), one of America's leading astronomers of the late nineteenth century,[19] took his readers on an imaginary flight through the universe to show how "we might overlook such an insignificant little body as our earth, even if we made a careful search for it.... The ancients had no conception of distances like this."[20] Newcomb's version of this myth did not assume that a large cosmos would imply that humans had no special role within it. That Newcomb did not make this further assumption may have had to do with his Christian upbringing and his respect for Christian scientific colleagues—despite Newcomb's own rejection of Christianity.[21] Newcomb's introductory astronomy text contains none of the other historical science-religion myths surveyed in this book.

Camille Flammarion (1842–1925), the most influential popularizer of astronomy internationally at the turn of the century,[22] outperformed Carey and Newcomb with a far more dramatic and antitheistic version of the "premodern small cosmos" myth. In a 1904 book (translated from the 1903 French original), Flammarion called astronomy "the science *par excellence*," one that tells us "where and what we are." Without *modern* astronomy we would "still be penetrated with the naïve error that reduced the entire Universe to our minute globule, making out Humanity the goal of the Creation, and should have no exact notion of the immense reality."[23]

The historical claim—that humans did not know the cosmos was large compared to earth's size until modern times—is demonstrably false. The philosophical claim—that the alleged discovery of a vast universe undermined theism—is just that: a philosophical add-on to an otherwise neutral scientific fact of cosmic immensity. It is not even clear philosophically or theologically how immense cosmic space would necessarily imply a lack of a divine plan for humanity.

Flammarion preached a new spirituality without traditional religion and without the concept of an exceptional moral status for human beings. The infinitude of space swallows us up in unlimited delight (or something like that).

> The reality is far beyond all dreams, beyond the most fantastic imagination. The most fairy-like transformations of our theaters, the most resplendent pageants of our military reviews, the most sumptuous marvels on which the human race can pride itself— all that we admire, all that we envy on the earth—is as nothing compared with the unheard-of wonders scattered through Infinitude.[24]

The last paragraph in Flammarion's book is a cosmic call to eternal life without theistic religion:

> As our planet is only a province of the Infinite Heavens, so our actual existence is only a stage in Eternal Life. Astronomy, by giving us wings, conducts us to the sanctuary of truth. The specter of death has departed from our Heaven. The beams of every star shed a ray of hope into our hearts. On each sphere Nature chants the paean of Life Eternal.[25]

How this amounts to eternal life for the individual human is left vague, but God is definitely not involved. Physicists in the late nineteenth century began talking about the inevitable thermodynamic heat death of the universe. This scientific discovery eventually undermined the godless cosmic eternal life that Flammarion had popularized. The cosmos, left to its own rules, would eventually reach a thermodynamic equilibrium in which no life is possible. The universe is living under a death sentence.

A Long Funeral for the Big Myth

The witty English controversialist G. K. Chesterton, a few years *prior* to Flammarion's international "big myth" bestseller, critiqued the "contemptible notion" that cosmic immensity "ought to over-awe the spiritual dogma of man." He wrote: "Why should a man surrender his dignity to the solar system any more than to a whale? If mere size proves that man is not the image of God, then a whale may be the image of God.... It is quite futile to argue that man is small compared to the cosmos; for man was always small compared to the nearest tree."[26]

Chesterton made the same point that Kepler did centuries earlier. Kepler had terminated his discussion of this topic with: "Yes, the crocodile or the elephant would be closer to God's heart than man, because these animals surpass the human being in size." Apparently neither Flammarion (1904) nor Bill Nye grasped this point.

Like Kepler and Chesterton before him, C. S. Lewis took on the big myth. In particular, his 1943 essay "Dogma and the Universe" debunked the myth.[27] After rehearsing how premodern intellectuals in the Western tradition had long embraced a point-like earth in a vast cosmos, Lewis identified the "real question." That is "why the spatial insignificance of the earth, after being known for centuries, should suddenly in the last century have become an argument against Christianity."[28] Although my research shows that the big myth probably originated in the nineteenth century, as Lewis suspected, its anti-Christian shape seems to have evolved gradually (not suddenly as Lewis surmised), and it did so from earlier Catholic-critiquing discourse.

So much for the historical aspect of the big myth. How about its philosophical pretensions? Lewis set up his critique with a forensic analogy: "When the doctor at a postmortem diagnoses poison, pointing to the state of the dead man's organs, his argument is rational because he has a clear idea of that opposite state in which the organs would have been found if no poison were present."[29] Similarly, if we try to *disprove* God by pointing out how small we are in a huge cosmos, we should clearly identify the kind of universe that is expected if God *did* exist. But Lewis argued that we fail here because of how human perceptions commonly operate:

> Whatever space may be in itself…we certainly perceive it as three-dimensional, and to three-dimensional space we can conceive no boundaries. By the very forms of our perceptions, therefore, we must feel as if we lived somewhere in infinite space. If we discovered no objects in this infinite space except those which are of use to man (our own sun and moon), then this vast emptiness would certainly be used as a strong argument against the existence of God.

Lewis then ran through the options from this point:

> If we discover other bodies, they must be habitable or uninhabitable: and the odd thing is that both these hypotheses are used as grounds for rejecting Christianity. If the universe is teeming with life, this, we are told, reduces to absurdity the Christian claim—or what is thought to be the Christian claim—that man is unique, and the Christian doctrine that to this one planet

God came down and was incarnate for us men and our salvation. If, on the other hand, the earth is really unique, then that proves that life is only an accidental by-product in the universe, and so again disproves our religion.[30]

Lewis dismantled anti-theistic rhetoric by walking us through the alternatives (including the existence or nonexistence of ET) and showing how atheists might find just about anything as a basis for disbelief. Turning to an aesthetic and emotive response to the subject of cosmic size, Lewis declared that if the universe

> were small enough to be cozy, it would not be big enough to be sublime. If it is large enough for us to stretch our spiritual limbs in, it must be large enough to baffle us. Cramped or terrified, we must, in any conceivable world, be one or the other. I prefer terror. I should be suffocated in a universe that I could see to the end of.[31]

Whether in a big or small cosmos, creatures like us will find emotional and aesthetic advantages and disadvantages. Lewis expressed his own preference for a gigantic universe, but neither size scenario grounds a good argument for human significance or insignificance, or for God's existence or nonexistence.

In a recent essay in a top-tier philosophy journal, *Noûs*, antitheist Oxford philosopher Guy Kahane agrees with Lewis that the size of the universe, whether big or small, has no substantial bearing on the question of God's existence or human significance. He shows that "facts about the dimensions of the universe seem…axiologically irrelevant." He means that cosmic size has no legitimate bearing on the value of anything within the universe. Kahane argues that significance relates to value: "For something to be important, it needs to possess, or bring about, enough value to make a difference—and that if something *is* important, then it merits our attention and concern" (i.e., it is significant).[32] Kahane agrees with Pascal, according to whom, even if (or when, given our knowledge of thermodynamics since Pascal) the cosmos destroys humanity and all life, "man would still be more noble than that which killed him, because he knows that he dies and the advantage which the universe has over him; the universe knows nothing of this." Kahane approvingly notes that in this argument for human significance, "Pascal actually focuses, not simply on our rational capacities, but precisely on our capacity to adopt the *cosmic standpoint*—to form a view of everything."[33]

The German-British philosopher Michael Hauskeller agrees with Kahane and Pascal that cosmic "size does not matter in itself" for human significance. But *larger* generally means *more powerful*, argues Hauskeller:

> My point is that perhaps our concerns are not so much about *size* as such as they are about *power*. While what is bigger is not necessarily more powerful than what is smaller, size is usually a good indicator of power, and if the size of a thing is so immense that there is not really a common measure between it and us, then it would be foolish to think we might stand a chance to overpower it in a confrontation.

So our feelings of insignificance arise not just from recognizing how small we are compared to the cosmos but also

> because we correctly infer from its immensity that we are utterly powerless to influence the course of (cosmic) events. Whatever we do, things will stay pretty much the same, and in the end, once we are all gone (and gone we will be), things will be exactly as they would have been if we had never existed at all.[34]

This might be an accurate depiction of the world *if* God does not exist, even as it underappreciates the stature of human intelligent causation compared to unintelligent natural causes—a point Pascal made effectively and put to Christian apologetic work. If, contrary to Hauskeller's belief, the biblical God exists, then there is eternal life for the redeemed (gone we will *not* be). Furthermore, the vast causal powers of the universe would be merely one tiny expression of the unlimited power of God as he upholds the world's existence.

Nicene Christian theology formulated in the fourth century well expressed the Judeo-Christian integrated grasp of *power, size, and significance*—divine, cosmic, and human. For example, Hilary of Poitiers (ca. 300–368) argued in his book on the Trinity:

It is the Father to whom all existence owes its origin. In Christ and through Christ he is the Source of all. In contrast to all else he is self-existent. He does not draw his being from without, but possesses it from himself and in himself. He is infinite, for nothing contains him and he contains all things; he is eternally unconditioned by space, for he is illimitable.... let our thoughts of the Father be at one with the thoughts of the Son, the only faithful witness who reveals him to us.

The vast cosmos comes from the Father's plan enacted through Christ in creating and sustaining all things. This self-existent, all-sustaining one revealed himself to ancient Jews in these seemingly paradoxical terms: "For thus says the One who is high and lifted up, who inhabits eternity, whose name is Holy: 'I dwell in the high and holy place, and also with him who is of a contrite and lowly spirit, to revive the spirit of the lowly, and to revive the heart of the contrite.'" God's omnipotent sustaining power expressed in cosmic immensity is interlocked with his loving power to enter into the lowly lives of those tiny image-bearing humans living in dependence on him. Nicene Christianity, explicating the New Testament in conversation with the Old, emphasized the incarnation of the eternal Christ as the perfect revelation of the all-encompassing omnipotent God, come to save us from sin and in so doing make the believers' little lives *significant permanently*—thus disclosing fully what Isaiah had cryptically intimated.

Even so, believers, similar to unbelievers, initially *feel* insignificant in the face of the cosmos because of its immensity and great causal powers. But the believers, with God's revelatory and redemptive help, realize that God has crowned us with honor that remains despite the daunting spatial and causal stature of the universe. That is the message of Psalm 8 ("When I look at your heavens...what is man that you are mindful of him?") as recognized by many prophets, philosophers, theologians, and astronomers over centuries of human awareness of cosmic immensity. Unbelievers are free to disagree, but the immense size and gargantuan causal footprint of the cosmos do not provide substantial grounds for rejecting God or for denying the significance of humans made in God's image. The history of ideas about cosmic immensity from Old Testament times to the present makes this point abundantly clear. Anti-theism must look for justification elsewhere.

How about the so-called Dark Ages, as an argument for the horrific and science-killing cultural damage Christianity caused? It is to this historical hypothesis that we now turn.

2

IDIOTS IN THE DARK

Had there been no Christianity, if after the fall of Rome atheism had pervaded the Western world, science would have developed earlier and be far more advanced than it is now.
    —*Biologist and New Atheist Jerry Coyne, "Did Christianity (and Other Religions) Promote the Rise of Science?"* Why Evolution Is True, *October 18, 2013*

The alleged precipitous decline of science in the "Christian" Middle Ages is an artifact of a cheap historical trick.
    —*Michael Shank, in* Newton's Apple and Other Myths About Science *(2015), 10*

In tempting the public to embrace prejudiced views against Europeans living between 400 and 1450, probably no one has done more than the astronomer Carl Sagan. In the more than five million printed copies of *Cosmos*, the companion book to his 1980 TV documentary series, which itself was seen by a half billion people, Sagan asked why the West slumbered "through a thousand years of darkness until Columbus and Copernicus."[1]

He blamed medieval Christianity. Sagan explained that "those afraid of the universe as it really is, those who pretend to nonexistent knowledge and envision a Cosmos centered on human beings will prefer the fleeting comforts of superstition."[2] Christianity, in his version of history, is a treacherous, superstitious science stopper. According to Sagan and so many other commentators, Christianity ushered in the "Dark Ages."

In his book, Sagan depicted these supposed Dark Ages with a graphic timeline that showed a gap in scientific advancement from the years 500 to 1500. The caption proclaimed, "The millennium gap in the middle of the diagram represents a poignant lost opportunity for the human species."[3] To this, historians of science David Lindberg and Michael Shank issued a stinging rebuke: "The timeline reflected not the state of knowledge in 1980 but Sagan's own 'poignant lost opportunity' to consult the library of Cornell University, where he taught."[4] There Sagan would have found a sizable literature documenting medieval scientific achievements.[5] These were not really "Dark Ages" at all.

For the description "Dark Ages" to make sense at all, the period must have been preceded by a time of illumination. So what I'll call the "dark myth" requires the powerful Roman Empire (in roughly the first four centuries AD) to have been characterized by scientific advancement. But contrary to the myth, Roman imperial prowess did not entail widespread scientific progress. With the exception of a small number of people centered in Alexandria, Egypt (chiefly Ptolemy and Galen writing in Greek), the vast empire—even Rome itself—was dominated culturally by people who cared little for science, other than what appeared in superficial handbooks and encyclopedias. Even those Romans who were sophisticated enough to read Greek rarely devoted themselves to the science written in that tongue. Michael Shank offers this comparative judgment: "In light of their resources and especially by comparison with the Islamic and Latin civilizations of the Middle Ages, the Romans engaged little with Greek science."

Shank explains that "Roman colonizers left Greek scientific works largely untranslated," meaning that those works "were mostly inaccessible when Latin alone defined literacy." He adds that "this situation was well entrenched by the time Constantine legalized Christianity in the fourth century," so that "when the Latin Christians expressed lukewarm attitudes toward Greek science, they were reflecting ambient culture, not changing it."[6]

Eventually educated Latin Christians *changed* the dying Roman culture they had inherited—salvaging some of it as they created a more science-friendly European culture. They did so, in large part, on the basis of the highly rational theological riches of the Judeo-Christian tradition. This is "how the West won," as sociologist Rodney Stark argues.[7]

The Dark Myth Rises

From where did the false idea of medieval Christians as "idiots in the dark," as we might frankly put it, derive? It began, without any anti-religious agenda, as mere chronological-stylistic snobbery. Some fourteenth- and fifteenth-century scholars in the humanities, especially literary types, initiated a vigorous campaign of declaring their own cultural superiority compared to their predecessors. They invented the notion of the Dark Ages between the lost political-literary glory of Rome and their own self-styled revitalization of it.

In the wake of the Reformation, Protestant historians exaggerated the image of a gloomy (Catholic) "Middle Ages." For example, the influential German Christoph Cellarius (1638–1707) commemorated the cultural "darkness" that the "New Era" initiated by Martin Luther overcame. David Lindberg and Michael Shank write, "By the nineteenth century, a layering of humanist, Protestant, and Enlightenment sensibilities had transformed Cellarius's tripartite division into the framework that historians from the European colonizing nations routinely used to structure their understanding of the globe."[8] Three periods of world history emerged: ancient, medieval, and modern. It is a familiar U-shaped story, with the bottom segment of the "U" representing cultural darkness. This historical template does not do justice to Europe, let alone the rest of the globe.

The cultural historian Jacob Burckhardt, one of the leading craftsmen of the historical label "Renaissance" (which means "rebirth") for the first segment of history after the Middle Ages, notoriously claimed in 1860 that the medieval mentality was characterized by blind adherence to "faith, illusion, and childish prepossession."[9] This is the Christianity-bashing characterization of the Middle Ages that Carl Sagan echoed. It is an untenable historical argument against the intellectual respectability of Christianity. Between the time of Burckhardt and Sagan, scholars documented many cultural accomplishments of the medieval period, including substantial achievements in the natural sciences. Although Burckhardt might be partially forgiven for his error, Sagan had no good excuse.

When did *scientists* begin weaving the so-called Dark Ages into their narratives? Sagan was hardly the first. I studied 130 English-language science textbooks published over four centuries, and the earliest I could find with the Dark Ages myth was published in 1809, and it was based on lectures that began in the 1780s.

Immersed in Ignorance

In Europe, the rhetoric of the French Enlightenment tilted the myth of the Dark Ages into a generic anti-Christian storyline during the eighteenth century. American science educators largely avoided that trend until the 1960s. The Dark Ages myth did pop up, but science educators used it to advance a specifically Catholic-bashing narrative as opposed to a broadly anti-Christian agenda.

The Presbyterian minister and educator John Ewing (1732–1802) retreated from Philadelphia in 1777 as the British army advanced on that great city. He returned with his family and was elected provost of the University of Pennsylvania in 1779. About this time Ewing began writing (and delivering over a twenty-year period) lectures on physics and astronomy that were published posthumously in 1809 under the revisionary editing of Robert Patterson, professor of mathematics in the same university. In Patterson's biography of Ewing, inserted at the beginning of this textbook, we learn colorful details about the man.

The Ewings "emigrated from Scotland at an early period of the settlement of our country."[10] After graduating from Princeton in 1755, John Ewing was "licensed to preach the gospel" and in 1759 became pastor of Philadelphia's First Presbyterian Church. Patterson reports that at this time Ewing chose "his profession," thinking "the study of theology congenial with his wishes, and calculated to permit him to mingle with it scientific researches."[11] From 1773 to 1775, with the consent of his congregation, he traveled to Britain to collect texts and raise support for American higher educational initiatives.

Ewing's pastoral successor at Philadelphia's First Presbyterian Church, John Blair Linn, in a funeral sermon excerpted in Patterson's biographical sketch, praised Ewing's command of "the learned languages," especially Hebrew. Linn went on: "Science was to him a powerful assistant in the labours of his sacred office. She was with him a handmaid to religion, and, aided by her, he was an able champion of the cross, both in the advocation of its cause, and in the repulsion of the attacks of impiety and error."[12]

Science as a "handmaid to religion" was, ironically, a slogan used to promote both science and the Christian religion during the Middle Ages.

Ewing introduced his physics and astronomy textbook with an alleged history of the battle for (and against) modern learning. "When learning began to flourish in Europe, it received a severe and long check from the desolations of war, and the irruptions of barbarous nations."[13] To be sure, repeated attacks by Germanic tribes disrupted learning in late antiquity and in the early medieval period. "But," Ewing insisted, "nothing so much prevented the propagation of knowledge, as the envious and illiberal craft of the heathen and popish priesthood."[14]

Ewing would have us believe that medieval Catholic authorities had "proscribed all books" that "might propagate real and useful knowledge" and that such books "were accordingly ordered to be…burnt, as it was deemed an unpardonable sin to read them." After Galileo's 1609 telescopic discoveries, Catholic priests supposedly continued this anti-intellectual mentality by forcing Galileo to "deny the truth of what he and many thousands had seen with their own eyes."[15] This is another false claim, as chapter 5 will demonstrate.

The Episcopalian minister-educator John Lauris Blake (1788–1857), like the Presbyterian John Ewing, blamed the Catholic Church for the alleged intellectual backwardness of the Middle Ages. Blake expressed the view in his *First Book in Astronomy Adapted to the Use of Common Schools* (1831). Although Blake's anti-Catholic historical views are more clearly stated in his *Conversations on the Evidences of Christianity* (also crafted for public school instruction),[16] his polemical Protestant bent colored the historical content of his science textbook. Referring to Copernicus's cultural context, Blake asserted: "Europe, however, was still immersed in ignorance, and the general ideas of the world were not able to keep pace with those of a refined philosopher. This caused Copernicus to have a few abettors and many opponents."[17]

Blake ignored the substantive contributions to astronomy and other sciences by medieval and early modern Catholics, including Copernicus himself—an administrative employee in the Catholic Church just below his bishop.

Chemist (and later historian) John William Draper (1811–1882) might appear to be another American science educator who attacked religion. His widely read *History of the Conflict Between Religion and Science* (1874) has long had the reputation of being one long complaint against anti-scientific Christianity. Yet even this work partially allies itself with Protestantism, if only in a weak, liberal sense.[18] More to the point, Draper wrote this tirade against Christianity as the great science stopper long after he had ceased his work in science education. In the 1840s and 1850s he had authored textbooks on chemistry, physical science, and physiology, but by the 1870s he was writing as a historian rather than as a science educator. The theme of "warfare" between science and religion remained largely absent from his earlier science textbooks.[19] Only as an elderly polemical historian did he transition himself—and further transition America—to the darkest version of the dark myth.

In Science Textbooks, the Dark Myth Gets Darker

In my sample of 130 textbooks, the earliest example of an anti-Christianity (as opposed to merely anti-Catholic) version of the Dark Ages myth occurs in the 1960s. It appears in UCLA astronomer George Abell's textbook that (as mentioned in chapter 1) also perpetuated the myth of premodern cosmic smallness. Here is how Abell spun the dark myth: "In medieval Europe no new astronomical investigations of importance were made. Rather than turning to scientific inquiry, the medieval mind rested in acceptance of authority and absolute dogma."[20]

Other prominent American astronomers have followed Abell's Dark Ages script. Donald Menzel and his coauthors wrote that "the lamp of science all but went out in medieval Europe." These were "dark times" during which people were more occupied with "war, art, letters, philosophy, and religion." But then the Renaissance came to the rescue, spurring an "awakening of scientific interest in western

Europe."[21] Similarly, Nicholas Pananides and Thomas Arny wrote that during the early Middle Ages "the acquisition of knowledge declined steadily because of the hostility that existed between the pagans and the Christians." They even claimed that "in their enthusiasm for orthodoxy, the Christians destroyed many of the pagan institutions." Consequently, "astronomy went into a state of dormancy."[22] But in reality, pagan institutions related to science had long been in decline before Christianity became a majority religion (as we will see in chapter 9).

In an astronomy textbook currently in use, Adam Frank, who calls himself an "evangelist of science,"[23] preaches, "In western Europe, the Dark Ages had descended and learning slowed to a crawl."[24] By Frank's lights, the previous enlightened era of "Greek genius" was characterized by "the demand to know the world through reason and mathematics and its rejection of the supernatural." Allegedly this Greek mentality (after the millennial Christian "Dark Ages" tried to undo it) "changed all of human history and led directly to the flourishing of science of our age."[25]

It is hard to know where to begin correcting such fanciful misconceptions about the Middle Ages. Contrary to Frank's assumption that the "rejection of the supernatural" enabled humans "to know the world through reason and mathematics," belief in the supernatural creation of the cosmos gave humanity solid conceptual ground for expecting to discover mathematically elegant natural laws issued from the maximally rational divine lawgiver.

The Dark Myth Challenged by Early Medieval Light: 400–1100

Some reserve the pejorative term Dark Ages for only the *early* Middle Ages (400–1100). Yet even then we see the light of learning.

Saint Augustine (354–430) contributed to Aristotelian physics in his *Literal Commentary on Genesis*.[26] More broadly, Augustine expressed confidence in our ability to read (discover) the "book of nature" because it is the "production of the Creator."[27] He insisted that we should proceed "by most certain reasoning or experience" to discern the most likely way that God established "the natures of things"—a popular medieval book title for works that emulated Augustine's investigative approach to the natural world.[28] He also noted that in any given case, multiple competing explanations of natural phenomena are possible, so discernment is needed. Interpreting the Bible and its relationship to natural knowledge (science) calls for similar caution, he advised. Indeed, Augustine affirmed that God wrote two books: the Bible and the cosmos. Both are intelligible to the human mind, and both require careful interpretation. Many other scholars and productive scientists through the Middle Ages and into modernity used this "two books" metaphor to affirm the harmony of Christianity and science.

Alongside Augustine, another highly educated Christian in the fading Roman world left a similar legacy for Europeans. The Roman senator and intellectual Anicius Manlius Severinus Boethius (ca. 475–524) accepted the earlier "proofs of the astronomers" of cosmic immensity (and Earth's relative tiny stature in this cosmic perspective).[29] Sitting on death row for an alleged conspiracy against the Ostrogothic leader Theodoric, who had displaced the last Roman emperor, Boethius found comfort in the smallness of his own suffering—and the relative insignificance of misapplied Ostrogothic authority—in the grand scheme of God's cosmos and God's loving universal rule. He synthesized Plato's philosophy with Christian theology to acknowledge the ideal of wise governance on earth that reflects God's rule over the universe.

In what (to our ears) sounds like rap music, we encounter some of the Christian roots of the idea of "laws of nature," which is so important to science:

> Creator of the starry sphere,
> Seated upon your timeless chair,
> You move the sky in swift gyration,
> Ordering with law each constellation.[30]

Boethius celebrated God's ordering of the four earthly seasons later in this poem, as he did in this subsequent poem:

> To each season God assigns
> Duties meet for his designs;

He prohibits changes made
To the sequence he has laid.
Headlong rupture of that chain
Unhappy outcome will obtain.[31]

Inanimate nature necessarily obeys God's rules—what we call natural laws. Humans, however, have a choice. Ostrogothic Rome tragically fell short of God's wise rule over all things, Boethius suggested. In Latin, and later in several vernacular European languages, Boethius's *Consolation of Philosophy* transmitted moral lessons alongside foundational scientific concepts through the Middle Ages.

Let us leave continental Europe for a moment. While many Irish scholars in the seventh and eighth centuries wrote on *computus* (astronomical calculations for determining the date of Easter), the English monk Bede (673–735) addressed astronomical and cosmological theory in the tradition of Ptolemy and Augustine. Historian Bruce Eastwood has summarized the importance of Bede's book *The Nature of Things* (ca. 701):

> Bede's work on the "natures" (the four elements and their subordinate parts) created by God represented a new stage in early-medieval Christian cosmology, informed by Augustinian categories and classical contents in extended detail. It became a model for a purely physical description of the results of divine creation, devoid of allegorical interpretation, and using the accumulated teachings of the past, both Christian and pagan.[32]

Note how Bede's Christian worldview was compatible with analysis of the natural world as a coherent system of natural causes and effects.

The Dark Myth Vanquished by High Noon Light: 1100–1450

About the year 1100, a tipping point came in the study of scientific argumentation and logic. European intellectuals graduated from limited translations and commentaries on Aristotle to a more extensive recovery of Aristotelian logic. As refined within a Christian worldview, this advance included reasoning methods well suited to natural science. Scholars called this form of argument "*ratio*" (reason), contrasting it with mathematical demonstration. *Ratio* uses premises inferred as likely true from sensory experience and then reasons from those starting points to probable conclusions.[33] Mathematics, by contrast, begins with first principles thought to be certain and then deduces conclusions that carry the same certainty.

A logic appropriate to science, *ratio* especially enriched the study of motion and change in the natural world. Historian Walter Laird writes:

> The study of motion in the Middle Ages, then, was not a slavish and sterile commentary on the words of Aristotle, but neither was it a failed attempt at the experimental science of motion that Galileo and Newton would establish in the seventeenth century. Rather, from a critical examination of the best sources available at the time, medieval logicians, philosophers, and theologians undertook to explain motion and change and the many puzzles surrounding them, developing in the process a series of new insights and analytic techniques that yielded a number of notable results. Part of the measure of their success—but only part—is that some of these insights and results had to be rediscovered later by Galileo and others in the course of the Scientific Revolution.[34]

Galileo either lacked awareness of or failed to acknowledge his work's medieval heritage, giving the false impression of a radical break from the past. That break has, since at least the 1830s, been called the "Scientific Revolution." But Galileo, a brilliant and often innovative scientist, stood on the shoulders of medieval giants.

The invention of the university had much to do with the rise of intellectual giants in the Middle Ages, beginning with the University of Bologna in 1088, followed by Paris and Oxford before 1200 and more than fifty others by 1450. The papacy supported this unprecedented intellectual ferment. Famous masters all over Europe attracted students into groups that became formal organizations called "universities," which could move to another town if local legal-economic conditions became less favorable. For example, an exodus from Oxford helped create the University of Cambridge in the early thirteenth century.

Universities provided additional stimulus to the medieval translation movement already under way, in which Greek and Arabic texts were rendered in the common European intellectual tongue of Latin. This movement greatly outperformed the comparative trickle of imperial Roman translations. If European Christians had been closed-minded to the earlier work of pagans, as the dark myth alleges, then what explains this ferocious appetite for translations? Many of the original Greek works and later Arabic commentaries that entered Europe in readable Latin concerned the knowledge of nature.

The proliferation of translations led to significant scientific advances. Consider the science of light and vision. The Franciscan cleric and university scholar Roger Bacon (ca. 1220–1292) read much of the newly translated work of earlier Greek and Islamic investigators, including Euclid, Ptolemy, and Ibn al-Haytham, or Alhazen (ca. 965–1040). By evaluating this past work and introducing some controlled observations—what we now call experiments—Bacon brought the science of light to its most sophisticated stage of medieval development.[35] Subsequent medieval and early modern authors summarized and reevaluated Bacon's work, transmitting it through books used in university instruction. That is how it came to the genius Johannes Kepler, whose account "helped spur the shift in analytic focus that eventually led to modern optics," in the words of historian A. Mark Smith.[36]

The medieval translation movement greatly affected universities. By one estimate, 30 percent of the medieval university liberal arts curriculum addressed roughly what we call science (including mathematics).[37] Between 1200 and 1450, hundreds of thousands of university students studied Greco-Arabic-Latin science, medicine, and mathematics—as progressively digested and improved by generations of European university faculty.

Michael Shank highlights the irony of all this for the dark myth:

If the medieval church had intended to discourage or suppress science, it certainly made a colossal mistake in tolerating—to say nothing of supporting—the university. In this new institution, Greco-Arabic science and medicine for the first time found

a permanent home, one that—with various ups and downs—
science has retained to this day. Dozens of universities introduced
large numbers of students to Euclidean geometry, optics, the
problems of generation and reproduction, the rudiments of
astronomy, and arguments for the sphericity of the earth.[38]

Were these universities secular or Christian? The question is
misleading because it imposes our own conceptual categories in
ways that distort the history of the university. Most medieval university faculty conceived of the world in a distinctly Christian manner, probably because they thought that this best explained reality.
Nobody forced them into this. Ecclesiastical authorities had very little influence on university curricula.[39] Universities were largely self-
regulating entities, but many university faculty and students enjoyed
special legal privileges the Church granted that local townspeople
resented. For example, in many university communities, members
accused of crimes would be tried in the more lenient ecclesiastical
courts rather than in the austere civil courts. Although generally
not composed of priests or monks, university communities typically
received generous support by means of a clerical status that carried
virtually no ecclesiastical responsibilities.

What about the 70 percent of university general education that
addressed nonscientific topics? Was that mostly theology? No. Few
students were even allowed to study theology. That privilege was
reserved for the select few who met the advanced educational requirements. Moreover, up through the thirteenth century, few universities
even had theological faculty. In the final century and a half of the
Middle Ages (up to 1450), the papacy approved the creation of more
faculties of theology. But still, of the main three higher faculties—
theology, medicine, and law—the last attracted the most students,
because in the growing bureaucracies of church and state, many more
jobs called for a law degree.

Note where the Catholic hierarchy spent a significant chunk of
its wealth between 1200 and 1800. Berkeley professor emeritus John
Heilbron opens his book *The Sun in the Church* by announcing, "The
Roman Catholic Church gave more financial and social support to
the study of astronomy for over six centuries, from the recovery of
ancient learning during the late Middle Ages into the Enlighten-

ment, than any other, and, probably, all other, institutions."[40] Michael Shank goes further: "Heilbron's point can be generalized far beyond astronomy."[41]

Looking Back with New Perspective

Although the Dark Ages myth began as an expression of chronological-stylistic snobbery by literary personalities who typically cared little about science, it soon morphed into anti-Catholic discourse among Protestant educators (including scientists). It finally became a generic assault on the intellectual respectability of Christianity as a whole. Interestingly, this final and darkest version of the dark myth did not become prominent in American science education until the sexual (not scientific) revolution of the 1960s. Might the past sometimes get reinvented partly for the sake of modern passions?

One does not have to be a Christian apologist to recognize the myth of the Dark Ages. The self-proclaimed atheist Tim O'Neill, who runs the blog *History for Atheists*, writes:

> The concept of "the Dark Ages" is central to several key elements in New Atheist Bad History. One of the primary myths most beloved by many New Atheists is the one whereby Christianity violently suppressed ancient Greco-Roman learning, destroyed an ancient intellectual culture based on pure reason and retarded a nascent scientific and technological revolution, thus plunging Europe into a one thousand year "dark age" which was only relieved by the glorious dawn of "the Renaissance." Like most New Atheist Bad History, it's a commonly held and popularly believed set of ideas that has its origin in polemicists of the eighteenth and nineteenth centuries but which has been rejected by more recent historians.[42]

Contrary to the dark myth, medieval European Christians culti-vated the idea of "laws of nature," a logic friendly to science, the sci-ence of motion, human dissection, vision-light theories, mathematical analysis of nature, and the superiority of reason and observational experience (sometimes even experiment) over authority in the task of explaining nature. Moreover, medieval innovators invented self-governing universities, eyeglasses, towering cathedrals with stained glass, and much more—not bad for people who have the unfortunate and undeserved modern reputation of being nearsighted idiots stumbling in the dark. Although any attempt to meaningfully label *any* age with a single descriptor is problematic, the so-called Dark Ages might be better labeled an "Age of Illumination" or even an "Age of Reason."[43]

3

FLAT EARTHERS

Many church leaders thought that ideas such as the sphericity
[roundness] of the Earth contradicted descriptions of the universe found in the Scriptures.
  —*John D. Fix*, Astronomy: Journey to the Cosmic Frontier
  (2011), 58

The erroneous [flat-earth] belief persisted through century after
century before the doctrine of a globular earth was fully established. Final doubt was swept away by the famous voyage of
Magellan [in 1522], one of whose ships first circumnavigated
the globe.
  —*David P. Todd*, A New Astronomy *(1897), 77*

Greek knowledge of sphericity never faded, and all major medieval scholars accepted the earth's roundness as an established
fact of cosmology.
  —*Stephen Jay Gould, "The Late Birth of a Flat Earth," in*
  Dinosaur in a Haystack *(1995), 42*

Celebrity astrophysicist Neil deGrasse Tyson has rightly opposed
the wacky recent resurgence of flat-earth belief,[1] but at the cost of
making a crucial historical error. He responded to flat-earth-promot-
ing rapper B.o.B in a January 25, 2016, tweet: "Duude—to be clear:

Being five centuries regressed in your reasoning doesn't mean we all can't still like your music." Tyson follower Andy Teal piped up three days later: "Five centuries? I believe the knowledge of Earth's shape goes back a bit farther than that." Tyson responded: "Yes. Ancient Greece—inferred from Earth's shadow during Lunar Eclipses. But it was lost to the Dark Ages."[2]

People stopped believing in a spherical earth during the Middle Ages? Not really. Medieval intellectuals in the Western tradition had many evidence-based reasons for holding to the earth's roundness. Those reasons included the curved shadow of the earth projected on the moon during a lunar eclipse. To deny medieval belief in a round earth is to be guilty of what I call the *flat myth*. This is the most enduring and pernicious component of the *dark myth* (chapter 2), and so it deserves its own chapter.

On January 26, 2016, Tyson's nephew Stephen Tyson, with Neil's backing and cooperation, rapped a response to flat earther B.o.B. The rap begins with Uncle Neil's introductory voiceover: "Flat Earth is a problem only when people in charge think that way. No law stops you from regressively basking in it." The intended lesson here is to avoid regressing to the Dark Ages when, supposedly, anti-science church folks were in charge. The song features Tyson, as a protégé of Sagan, confronting B.o.B:

> He learned the game from Carl Sagan, you can never check him
> You say the Earth is flat and then you try to disrespect him?
> I'm bringing facts to combat a silly theory
> Because B-O-B has gotta know the planet is a sphere G . . .
> And he's a Mason cuz a brotha's gettin' paid?
> When the ignorance you're spittin' helps to keep people enslaved?
> I mean mentally.[3]

Neil deGrasse Tyson has indeed learned the Carl Sagan game of dissing Christianity with the Dark Ages myth. The Christians in charge of these horrible ages are alleged to have enslaved humanity ("I mean mentally") with flat-earth nonsense. (Incidentally, Europe largely extinguished slavery during the Middle Ages—a major triumph of biblically informed reason and conviction over extensive Roman slavery.)[4] Neil's ending monologue to his nephew's rap repeated his previous day's tweet about B.o.B's "being five centuries regressed" for believing in a flat earth.

Tyson is obviously right about how ridiculous contemporary flat-earth belief is. Some "believers" such as Shaquille O'Neal and Kyrie Irving of NBA fame later said they were only joking or trolling. And who can tell what the small number of people behind today's flat-earth societies *actually* think? If most of them are joking or trolling, it would come as little surprise.

But the fact is that Tyson, probably the world's most influential public voice for science, is spreading misinformation about medieval views of the earth's shape.[5]

Many of Tyson's misconceptions on the subject trace back to Sagan and numerous other astronomers and popular writers. Eventually these lines of influence go back to nineteenth-century writers such as the chemist-historian John William Draper, introduced in the previous chapter. Draper claimed that medieval Christians believed "the Scriptures contain the sum of all knowledge." They therefore "discouraged any investigation of Nature" (a falsehood addressed in my previous chapter), including the scientific study of the earth's shape. Supposedly this "indifference continued until the close of the fifteenth century," when nonscientific commercial motivation to explore the world finally settled "the question of the shape of the earth...by three sailors, Columbus, De Gama [sic], and, above all, by Ferdinand Magellan."[6]

Columbus Proved Earth Is Round?

Of the 1,200 American college students I have taught astronomy over the past quarter century, the vast majority learned from their precollege teachers that Europeans in the Middle Ages were ignorant of earth's roundness until Christopher Columbus proved it in 1492. I know this because I routinely asked for a show of hands.[7] Although only some of my students had previously detected the typical anti-Christian slant to the story, they quickly grasped how this fake history perpetuates the myth of warfare between science and Christianity.

The earliest instance that I have found of the false claim that Columbus proved earth's sphericity occurs in an 1818 British-

American juvenile astronomy textbook. Listen for the note of Protestant schoolboy anti-Catholic polemic:

> Happily, however, for mankind, the discoveries of a Columbus silenced the senseless clamour of the Romish clergy; and the figure of our globe, together with its motions annual and diurnal, are now explained in so clear and perspicuous a manner, that a competent knowledge of them, may be attained with the utmost facility by every attentive reader.[8]

Washington Irving's widely read *History of the Life and Voyages of Christopher Columbus* (1828) claimed that a group of scholars in Salamanca opposed Columbus's plea for royal finances for his trip *westward* to reach the *East* Indies—partly on biblical grounds: "To his simplest proposition, the spherical form of the earth, were opposed figurative texts of Scripture." Irving suggested that the theologians in the group misinterpreted biblical texts to infer that the earth "must be flat." Irving's historical assessment misses wide of the mark. From the time of Augustine, virtually no Western Christian scholars made biblical arguments for a flat earth. And yet Irving claimed that Columbus "was in danger of being convicted not merely of error, but of heterodoxy."[9] This was simply not so.

Irving suggested that others at this Salamanca meeting who were "more versed in science admitted the globular form of the earth." But, he wrote, they launched yet other arguments for why the proposed trip would fail. The objections mentioned included "insupportable heat" near earth's equator and the opinion "that the circumference of the earth must be so great as to require at least three years to the voyage," making it seemingly impossible to bring sufficient provisions.[10] This last objection was based on an approximately correct view of earth's circumference, which Columbus woefully underestimated owing to his greater reliance on sensational explorer reports and marginal outdated science rather than current majority science.[11] Fortunately, Columbus and his crew encountered the New World and so did not perish.

Irving inaccurately depicted Columbus as the victor over absurd theological objections to a journey that would usher in scientific and geographical modernity:

We must not suppose, however, because the objections here cited are all which remain on record, that they are all which were advanced; these only have been perpetuated on account of their superior absurdity. They were probably advanced by but few, and those persons immersed in theological studies, in cloistered retirement; where the erroneous opinions derived from books, had little opportunity of being corrected by the experience of the day.[12]

Irving blended historical fact and fiction in his excessively heroic tale. The real issue debated at the time of Columbus's voyage was not the earth's shape but its size (and the size of oceans in relation to land). There was no significant medieval tradition of arguing theologically or scientifically for a flat earth. Accordingly, Columbus's crew was not worried about sailing off the world's edge.

Historian Rudolf Simek summarizes the growth of academic and popular late medieval knowledge (including that of sailors) of earth's roundness. This awareness continued up through Columbus's time:

From the 12th century on astronomy handbooks and university teaching led to the classical Ptolemaic view of the world being made known firstly to most clerics and subsequently to all university students. By the 13th century the spherical shape of the earth, and therefore also its theoretical circumnavigation, had found its way not only into scholarly but also into popular literature. This included one of the most widely read travel books of the age, *Travels* by Sir John Mandeville.[13]

George Darley's *Familiar Astronomy* (1830) provides another early instance of the Columbian version of the flat myth. Taking the form of a fictitious dialogue occurring over twelve evenings, the book spans the literary territory between a standard juvenile astronomy textbook and fiction.[14] John Taylor, "Bookseller and Publisher to the University of London," published *Familiar Astronomy*. The preface suggests that one can learn important truths through the fictional dialogue. The wisest character among the dialogue participants, Franklyn, says that in "ancient times when some of the wisest philosophers were more ignorant of terrestrial geography than a modern schoolboy, great mistakes were committed in astronomy. It was then supposed, that the Earth was nearly a perfect *flat* surface, the land portion of which was collected in the middle, while a surrounding ocean stretched illimitably to the sphere of the stars." The dialogue continues:

> How soon was the truth [of earth's roundness] discovered? said Eugenia.
>
> Why, about the exact time there is some obscurity. It is generally supposed that the true system of the world, according as it was suspected by Copernicus, the Prussian astronomer, in 1543, and demonstrated by Newton in 1687, had not been wholly unknown to Pythagoras, the Grecian sage, who flourished about 550 years before Christ. Various others, since that time, have, in various ways and degrees, approached the truth....
>
> But, until the age of Copernicus, men in general were...unacquainted with...the *cosmography* of the world, or, in other words, its true system of arrangement.... Before Columbus discovered America—or rather, before the first complete voyage round the Earth,—how could it be known to what extent the ocean spread itself? This, indeed, put the matter beyond doubt: navigators, by continually sailing forward in the same direction, found that they at length returned to the place whence they set out; and this it is which affords the only direct and practical proof of our Earth being a round body.[15]

Although many ancient cultures assumed the earth was flat, by the fourth century BC Greek thinkers had voiced a number of impressive arguments for earth's sphericity. During continued debate over earth's shape in the remaining pre-Christian centuries, the round earth enjoyed wide acceptance among intellectuals. Most early Christian scholars adopted the spherical view as well. This trend held among Christian theologians, artists, poets, scientists, and university-educated bureaucrats into the Middle Ages. Before Columbus, therefore, several hundred thousand university-educated Europeans had become acquainted with quite a few arguments for earth's roundness. Countless others in communication with these educated men, and conversant with popular literature outside the university, would likewise have known about earth's sphericity. So Darley's schoolboy tale about pre-Columbian "men in general" being "unacquainted" with such truths is false.

The Flat Myth Goes to College

In my sample of 130 textbooks, the first college textbooks to indulge the flat myth appeared at the turn of the twentieth century. But eminent astronomers perpetuated the flat myth without embedding it within the myth of the Dark Ages. Amherst College's David Todd wrote in 1897 that flat-earth belief "persisted through century after century" until Ferdinand Magellan "swept away" all doubts by circumnavigating the globe in 1522.[16] Notice that Todd avoided Columbus (who had been made especially well suited for anticlerical rhetoric after Washington Irving's biographical extravaganza) and replaced him with Magellan.

In 1906 University of Chicago astronomer Forest Moulton offered no anticlerical slant while including Columbus in his telling of the flat myth: "It is an undoubted fact that there was no general acceptance of the idea of the globular form of the Earth until after Columbus was supposed to have sailed to India, going westward from Spain. The epoch of exploration and discovery which followed his voyages made the theory well known and caused it to be accepted throughout Europe."[17]

In another textbook, Carleton College's astronomer Edward Fath likewise repeated the flat myth without attacking Christianity.[18]

But in the 1960s, things changed. Adopting the anticlerical posture of Washington Irving and other nineteenth-century polemicists, textbook authors began to depict ancient and medieval Christians as exceedingly anti-intellectual about earth's shape, and more. Wellesley College astronomer D. Scott Birney alleged: "Church scholars refused to accept the notion that the earth was round. Anyone, they reasoned, could see that it was flat."[19] The latest edition of John Fix's *Astronomy: Journey to the Cosmic Frontier*, currently used in colleges, regurgitates the myth:

> Part of the reason for the loss of Greek astronomical knowledge can be attributed to the antagonism of the early Christian

church to many of the features of Greek astronomy. Many church leaders thought that ideas such as the sphericity of the Earth contradicted descriptions of the universe found in the Scriptures. Rather than accept the spherical shape of the Earth and the celestial sphere, some Christian scholars such as Lactantius and Kosmas argued that the Sun, after sunset, traveled around the horizon toward the north and then east to rise again in the morning. Not all of the Christian scholars were as ready to reject Greek astronomy.[20]

This is the only currently used college astronomy textbook I could find (using VitalSource.com as a portal to most contemporary college textbooks) that continues to perpetuate such a ridiculous story. Most astronomy textbook writers have finally caught up with several generations of historians of science.

A Round Earth in Late Antiquity and the Early Middle Ages

The vast majority of early church scholars did not argue for a flat earth—with a notable exception: a small group of Greek-speaking theologians centered, in the fourth and fifth centuries, in Syrian Antioch.[21] What explains this minority opinion?

Diodore of Tarsus, in his tract *Against Fate*, criticized the traditional Greek spherical heaven-earth idea as the pagan basis for deterministic astrology. Diodore tried to force the biblical metaphor of the "tent" appearance of the heavens into a nonspherical cosmological theory of the world, which included a flat earth. We know Diodore's tract, now lost, had little influence because it was forgotten until the prominent ninth-century church leader Photius of Constantinople mentioned it and declared that its "biblical" arguments for a flat earth lacked cogency.

Only a few others perpetuated Diodore's failed attempt to associate the Bible with a flat earth. Those include Theodore of Mopsuestia (ca. 350–428/29), John Chrysostom (ca. 347–407), and Severian of Gabala (d. after 408). The last, in his *Homilies on Genesis*, suggested a tabernacle-shaped cosmos with a flat earth at the bottom. But the widely respected Alexandrian Christian theologian-scientist John

Philoponus (ca. 490–570) repeatedly criticized Theodore and his tiny band for misuse of the Bible.[22]

Then there is Cosmas Indicopleustes's cosmological oddity, entitled *Christian Topography* (ca. 550). Cosmas lived mainly in Byzantine Egypt but had intellectual roots in the Syrian-Antiochian tradition. His book was the last ancient-medieval attempt in the Greek language to extensively argue that the Bible licenses a flat earth. We know his work was largely ignored because there are only two extant copies of it and the only medieval European who certainly read *Christian Topography* was the same Photius of Constantinople who argued against Diodore, the founding father of this failed tradition.[23]

Cosmas was rediscovered at the end of the seventeenth century and used to bolster the newly fashioned false story that people in the Middle Ages typically believed in a flat earth. But historian C. P. E. Nothaft, who knows the primary literature well, argues that "the polemical tone of his *Topography* suggests that . . . [the spherical earth] remained the dominant view in his own time, that is, even among Christians."[24]

Flat-earth belief was even rarer among Western Latin-speaking Christians. Lactantius (ca. 240–325) is the only scholar that flat-myth perpetuators use to spin their story that Christianity impeded scientific progress. But his case is a curious one. Although he identified himself as a Christian convert from paganism, he never accepted the deity of Christ and so, for this and other reasons, was condemned as a heretic after his death.[25] True, he tried to make the case that the Bible teaches the earth is flat. But for such manipulation of the biblical text and science illiteracy he was largely ignored or ridiculed.

Augustine, not Lactantius, established the standard Christian approach to questions like earth's shape. As a theologian, Augustine assumed that the received scientific consensus of his day was correct about earth's roundness. We see this in his unfinished book *On the Literal Interpretation of Genesis* (ca. 401–415). In this work he pondered the timing of the creation of "Day" and "Night" described in Genesis 1:3–5. Arguing against the hypothesis that the creation of one took place before the other, Augustine wrote that "during the time when it is night with us, the presence of light is illuminating those parts of the world past which the sun is returning from its setting to its rising, and that thus during the entire twenty-four hours, while it circles through its whole round, there is always day-time somewhere, night-time somewhere else."[26]

This argument works *only* if earth is round (in at least the east-west dimension); a flat disc or cube-shaped earth (or countless alternatives) would be incompatible. Augustine reasoned, based on a round earth receiving sunlight, that "Day" (as created and ordered by God) and "Night" (as the "privation" of light as ordered by God) would always be *simultaneously* present on either side of a half-illuminated globe. Note how astronomical reasoning about sphericity and sunlight assisted in a "literal" textual interpretation, as Augustine called it.

This, incidentally, undercuts a common misconception today that a "literal" reading of the Bible presents us with a flat (or stationary) earth. Augustine's approach amounts to saying that "literal" means "accurate" (as much as possible) as we take into account all we know (at a given time in history) about the biblical text, the world, and anything else that is relevant. Scholars like himself, he wrote, aim to "show that whatever we have been able to demonstrate from reliable sources about the world of nature is not contrary to our literature."[27]

Note also that Augustine's scientifically informed literal interpretation of Genesis works regardless of what is really moving: the earth or the sun. In fact, Augustine stressed the point that the Bible does not teach *any* cosmology (neither a flat earth nor a round earth). Neither will the Bible ever contradict any well-established scientific discovery. Many prominent figures in medieval and modern science have likewise held this view (as we will see later).

But Augustine also advised us in general to hold natural knowledge hypothetically (including even earth's sphericity). Why? Because our confidence in such propositions is lower than our confidence in the clear teachings of the Bible. (Of course, today earth's approximate roundness is beyond reasonable doubt.) Nothaft summarizes Augustine's perspective:

> One of the guiding concerns of Augustine's interpretation of Genesis was to avoid confrontation between the literal sense of Scripture and the knowledge of the natural world that had been attained by the philosophers. This explains his repeated assertion that quarrels over details only detract the Christian reader from the essential aim of biblical literature, namely his

personal salvation. In this context, his neutral presentation of the spherical model as "hypothetical" is best understood as part of his effort to liberate the exegesis of Genesis from the strictures imposed by the need to reconcile it with any particular cosmological theory.

Nothaft reminds us, too, that Augustine's discussions of the physical world "clearly reflect his classical or 'pagan' education, for which reason it is no surprise that some of his remarks presuppose the spherical conception of heaven and earth."[28]

Beyond this, Augustine cited Psalm 148 ("Praise the LORD from the heavens") to support the idea that intelligent beings should praise God by directing their intellect toward his creation. To study the cosmos can be more than vain curiosity; it can be a form of worship, he urged.[29] Indeed, throughout the Middle Ages and beyond, scholars praised God by studying the world, including the evidence that pointed to a round earth.

Round-Earth Arguments in the Late Middle Ages

Imagine the year is 1300. You are a student at the University of Salamanca, Spain's oldest university, founded in 1134. In class you have learned of Aristotle's argument for a spherical earth based on the changing positions of the constellations as one travels north or south. This is a standard part of the medieval curriculum, and you wish to demonstrate it for yourself. How will you go about this?

First you note that Polaris (the North Star, which appears motionless throughout each night) is located about 40 degrees above your horizon in Salamanca. Then you travel to Punta de Tarifa, the southernmost point of the Iberian Peninsula and continental Europe (where the Mediterranean Sea opens to the Atlantic Ocean). There you find that Polaris appears only about 35 degrees above the horizon (see Figure 3.1, next page). You notice another difference from Salamanca: in Punta de Tarifa, fewer stars are circumpolar (meaning they never go below the horizon, as illustrated in Figures 3.1 and 3.2). For example, the constellation Cassiopeia in Figure 3.2, in its apparent twenty-four-hour cycle of circling counterclockwise around

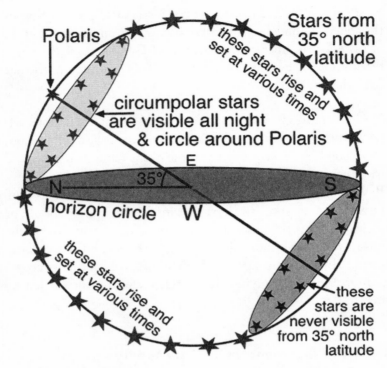

Figure 3.1: The stars you see at night depend on your latitude

the virtually stationary Polaris, dips partially below the horizon and is not always fully visible from Punta de Tarifa at night. Viewed from Salamanca, however, Cassiopeia is visible above the horizon throughout each night.

Almost every medieval university student learned a simple explanation for such facts: that earth was round (in at least the north-south direction). This and other observation-based arguments combined to present a very strong case for a spherical earth. The most used introductory university astronomy textbook from the mid-thirteenth to the end of the seventeenth century, *De sphaera* (On the Sphere) by Johannes of Sacrobosco, included a range of arguments for a spherical earth. For example, Sacrobosco noted that the rising time of a particular stellar constellation grew later each night as one travels west, providing evidence for an east-west curvature of the earth.[30]

Ironically, my students have typically been *less* able to defend earth's sphericity by such empirical arguments than the average medi-

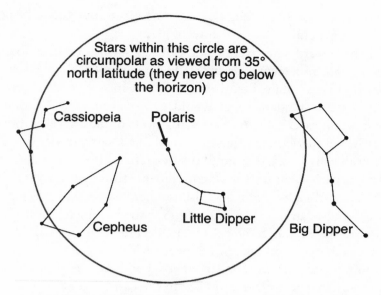

Figure 3.2: Stars visible throughout the night from 35 degrees north latitude

eval student. Upon completing my astronomy course, they finally caught up to the Middle Ages!

Most students today accept the roundness of earth by believing authoritative pronouncements from teachers rather than by reasoning from observations. That is their loss.

Full Circle, Then and Now

As we have seen, it is absurd to say that educated medieval people typically believed in a flat earth.

And yet the myth endures. Why?

Historian Jeffrey Burton Russell, the leading flat-myth buster, uncovered the chief motivation behind the *proliferation* (not the origin) of this tall tale. The myth gained traction in the late nineteenth century, for reasons having little to do with astronomy but much to do with evolution: "The reason for promoting both the specific lie about the sphericity of earth and the general lie that religion and science are in natural and eternal conflict in Western society is to defend Darwinism."[31] Of course, many Darwinists have not perpetuated the flat myth or have been religious in some sense, but Russell's point stands as a generalization worth recognizing.

In this regard, the twenty-first century seems not that different from the late nineteenth century. On January 21, 2009, I testified before the Texas State Board of Education in a successful plea for robust science standards that would give teachers the freedom to include scientific criticisms of certain aspects of modern evolutionary theory in the curriculum. Those opposed to allowing students to hear these criticisms paraded around in silly archaic clothing and distributed flyers with a painting of a Columbus-era ship tipping over the flat earth's edge. As a historian and philosopher of science, I was amused by the activists' antics. It was like a visit to the nineteenth century. The narrative of the activists' handout vaguely associated Darwin doubters (anyone who recognizes peer-reviewed scientific criticisms of certain aspects of evolutionary theory) with "flat earthers," a rhetorical ploy that Russell traced back to the earliest controversies over Darwin's theory in the 1860s.[32]

Today's most influential critics of Darwin encourage the teaching of scientific arguments both for and against the latest versions of evolutionary theory, thus emulating (in a revised, modern way) the medieval university practice of "disputation," in which students and faculty would openly debate controversial topics.[33] As a required step toward earning a bachelor of arts degree, medieval university students in their third and fourth years of study (often youngsters not much past sixteen years old) became active participants in disputations rather than just onlookers. Although only a minority of students completed the last two years of rigorous disputation, everyone was exposed to this method of critical thinking, at least as observers who could weigh the evidence in their own minds.[34] To fall short of something like these rigorous pedagogical techniques *today* would foster the very intellectual darkness popularly associated with the flat-earth myth.

Fried Bacon

I must confess I have held back a tasty little secret about the flat myth's early history. Although the myth became common in English-language juvenile astronomy textbooks in the early nineteenth century and proliferated in the late nineteenth century, especially as a rhetorical move against Darwin doubters, 1607 marks the earliest known instance of the myth.[35] It occurred in an unpublished work, *Cogitata et visa de interpretatione naturae* (Thoughts and Conclusions on the Interpretation of Nature), by the famous English science advocate and bureaucrat Francis Bacon. He wrote:

> Among the Greeks those who first suggested to men's untutored minds that thunder bolts and storms had natural causes were condemned for impiety. On the accusation of some of the early Christian fathers Cosmographers, who on clear evidence which no sane man could reject today, claimed that the earth was a sphere and therefore inhabited at the antipodes, fared little better than the Greeks. They were brought to trial for impiety.... In our own days discussions concerning nature have been subject to even harsher constraint.[36]

Bacon issued a revised version of this statement in his book *Novum Organum*, published in 1620.[37] Despite his dramatic claim, no major church council declared heretical either the roundness of earth or the antipodes. Antipodes referred to conjectured continents on the opposite side of the round earth. Did a race of rational creatures undescended from Adam and Eve live there? Had they sinned? Such questions had been entertained in ways similar to speculation today about extraterrestrial intelligence. These hypothetical considerations provided material for fascinating debates within Christendom. People realized that the earth could be round and yet lack rational creatures undescended from Adam and Eve.

Although there were no trials aimed at identifying round-earth or antipode advocates, my next chapter explores an unfortunate heresy trial that ended just seven years before Bacon, in his *Cogitata* manuscript, cooked up the flat myth. In 1600 the Roman Inquisition burned Giordano Bruno at the stake, in part, we often hear, because he speculated about inhabited extraterrestrial worlds. Did this make him a martyr for science? The apparently intense Protestant Christian Francis Bacon planted a seed for this view with his veiled comment "In our own days discussions concerning nature have been subject to even harsher constraint." Bacon appears to have conveniently packaged within a single paragraph the rudiments of two science-religion warfare myths that Protestants came to use in anti-Catholic polemics. In this way, Protestants unwittingly prepared the way for antitheists to launch a global assault on Christianity. Bruno became one of the icons of this culture war.

4

BURNING BRUNO

Bruno was burned at the stake—largely for supporting Copernicus, suggesting the sun to be a star, and suggesting that space was infinite.

—*Paul G. Hewitt,* Conceptual Physics *(1974), 1*

[Bruno] was sent to Rome to stand trial...for espousing a heliocentric and infinite universe.

—*Randy Moore et al.,* Chronology of the Evolution-Creationism Controversy *(2010), 13*

Bruno, the philosopher and mystic...was not executed for Copernicanism but for a series of theological heresies centering on his view of the Trinity, heresies for which Catholics had been executed before. He is not, as he has often been called, a martyr of science.

—*Thomas S. Kuhn,* The Copernican Revolution *(1957), 199*

Why did the Roman Catholic Inquisition burn Giordano Bruno at the stake in 1600?

Many people think he was executed because he believed in Copernican astronomy, extraterrestrial intelligence, and an infinite universe. In a widely shared conception of science and its history, this has made him a martyr for science. As we saw at the end of the previous chapter, Francis Bacon planted a seed for this interpretation of Bruno's death.

But Bruno's life and legacy are much more complicated than this tragic tale. On what basis did the Inquisition identify him as a heretic? By examining this episode, we will learn not only about Bruno but also about his confused interpreters over four and a half centuries. There is also much to learn about the contested religious implications of other worlds and other intelligent life beyond our own.

The Cultic Image of Bruno

On March 27, 2017, Giordano Bruno enthusiast Bruno Borges was reported missing from his family home in Brazil. For more than three weeks prior to his disappearance, this twenty-four-year-old Brazilian and his friends, bound by an oath of secrecy, had transformed his locked bedroom into a shrine to Giordano Bruno, the occult, and aliens. A $2,500 replica of the Bruno statue in Rome dominated the room. The walls, floors, and ceiling were adorned with thematic inscriptions, diagrams, paintings, and trinkets. Encrypted messages and books were also on display.

What happened to Borges? Social media buzzes with ideas. Some say aliens abducted him. Others maintain that Borges was a reincarnation of the sixteenth-century heretic, noting the striking resemblance between the Brazilian and the Italian Bruno. But the engraving of Giordano Bruno behind this claim was produced by a French artist for the frontispiece of an 1830 German publication of Bruno's collected works. We have little idea of what Giordano actually looked like.[1] Relatives say Borges, a psychology student, often requested funding for his secretive project, claiming that he was writing books that would "change humanity in a good way." Brazil's Criminal Investigation Department failed to locate Borges.[2]

That's how the latest Bruno story was left hanging for the vast majority of the English-speaking world. But, in fact, the Brazilian Bruno returned home after nearly five months, which was (conveniently) soon after some of his books began selling well. In a Google-translated story (the August 13, 2017, source is in Portuguese), Bruno Borges denied what the police suspected—that all this was merely a publicity stunt to sell his books on "cosmogonic theory," aliens, the occult, "theory of knowledge absorption," and other fantastic topics.[3] Giordano Bruno was Bruno Borges's mascot—a fitting choice, as we will see.

Giordano Bruno's statue in Rome, with its odd replica in Brazil, was originally installed in 1889 at the site of his execution. So we return to the scene of the crime. This is a good place to start digging for clues as to why Bruno was declared a martyr for science.

Our first witness is the chemist John William Draper, whom we met in chapter 2. In his *History of the Conflict Between Religion and Science* (1874), he anticipated the day "when posterity will offer an expiation for this great ecclesiastical crime." He thought a "statue of Bruno" would "be unveiled under the dome of St. Peter's at Rome." According to Draper, the statue would commemorate Bruno as one of "countless martyrs" for discovering religiously offensive scientific truth. Listen to Draper's memorial of Bruno:

> No accuser, no witness, no advocate is present, but the familiars of the Holy Office, clad in black are stealthily moving about.... He is simply told that he has brought upon himself strong suspicions of heresy, since he has said that there are other worlds than ours. He is asked if he will recant and abjure his error. He cannot and will not deny what he knows to be true.[4]

Bruno did believe in other inhabited worlds, but not for any compelling scientific reasons. And there was much more to the story than this—including his advocacy of aspects of ancient Egyptian religion updated by his own pantheistic philosophical theology. Contrary to Draper, there were *multiple* witnesses (especially his own books) to his wild ideas. But let us focus on Draper's correct anticipation of the 1889 Bruno statute as an emblem of free speech and scientific autonomy.

Soon after Italy's establishment as a secular state, a fraternity of Roman college students enlisted the help of eminent thinkers and leaders from around the world to raise a statue in Bruno's honor. Among the supporters was the German biologist Ernst Haeckel, now known for his inaccurate embryo drawings, often reproduced in textbooks, that offered spurious evidence for Darwinism. The inscription at the statue's base proclaims: "To Bruno, from the generation that he foresaw." The bronze Bruno faces north toward the Vatican with a penetrating glare shaded by his friar's hood. The annual commemoration on February 17 includes a wreath, flowers, poems, and candles at the feet of Bruno. But there is more, as a Bruno biographer explains:

> The Italian Association of Free Thinkers sets up a microphone at Bruno's feet; meanwhile, the atheists and the pantheists, carefully separated from each other, lay out their tables of books and leaflets on opposite sides of the piazza. In between them, the Free Thinkers guard their microphone jealously, wresting it in turn from the man in the sandwich board who claims to be Giordano Bruno incarnate.... This microphone is only for organizations, the Free Thinkers declare to all their competitors for the space beneath Giordano's lowering gaze: the Masons, the atheists, and the pantheists, all of whom claim the hooded friar as their very own spiritual leader.[5]

Draper's 1874 book, which was already in its nineteenth edition by 1885 (and translated in multiple languages), might have helped the Roman students succeed in their international subscription campaign for a Bruno statue. Camille Flammarion's *Astronomie populaire* (1880) probably helped the cause as well. In the preface to the first English translation of Flammarion's book, J. Ellard Gore notes the unprecedented success of the original French tome: "No fewer than one hundred thousand copies were sold in a few years—a sale probably unequalled among scientific books."[6] The anticlerical, and former Catholic, ET enthusiast Flammarion asserted that Bruno "was burnt alive at Rome in 1600 for his astronomical and religious opinions, and especially for his convinced affirmation of the doctrine of the plurality of worlds."[7] The "plurality of worlds" is a traditional way of referring to other possibly habitable celestial objects, or even other universes.

Textbook Mythology

In English-language astronomy textbooks, the earliest occurrence of the "Bruno as martyr for science" claim that I have found occurs in Herbert Howe's *Elements of Descriptive Astronomy* (1897). Howe implied scientific martyrdom when he wrote, "Giordano Bruno, who had unsparingly exposed many of the absurdities of the Aristotelian system of natural philosophy, had pointedly ridiculed it, and had propagated heretical notions (as, for instance, that the stars were suns), was burned at Rome."[8]

Harvard University's pioneering astronomer Cecilia Helena Payne-Gaposchkin (1900–1979), who experienced unfair treatment as a result of her sex, suggested in her 1954 textbook that Bruno was persecuted for his scientific support of Copernican theory: "At the end of the sixteenth century the theory of Copernicus was warmly, if not hotly, upheld by Giordano Bruno, who welcomed its aid in his attack on Aristotle. Bruno was imprisoned, excommunicated, and burned at the stake in 1600, and scientific men of succeeding generations cannot have been unmindful of his fate."[9]

This depiction of Bruno as a scientific martyr is a myth. Yet it lingers in textbooks, especially since Payne-Gaposchkin's telling of it. About a third of astronomy textbooks in use today tell this highly misleading story.

Were Bruno's Copernicanism and his belief in an infinity of inhabited worlds central to his trial by the Roman Inquisition? The first part of the question, about Copernicanism, is easily answered: no. Historians of science settled this years ago. Jole Shackelford explains that "the Catholic church did not impose thought control on astronomers, and even Galileo was free to believe what he wanted about the position and mobility of the earth, so long as he did not *teach* the Copernican hypothesis *as a truth* on which Holy Scripture had no bearing."[10] The second part of the question, regarding ET and many worlds, is much more complicated, and interesting. The needed documentation became widely available only recently, so we can now synthesize a fairly precise answer.

We begin by tracing the development of ideas about alien worlds in relation to the status of earth in the cosmos.

Places for ET to Live?

Ancient pagans in Greco-Roman civilization vigorously debated whether other inhabited worlds beyond our own exist. The atomists

believed in an infinitely large universe within which random interac-
tions among material atoms sometimes produced life, such as that on
earth. In the last five centuries BC, prominent voices for this opinion included Leucippus, Democritus, Epicurus, and Lucretius. Given
the unlimited probabilistic resources their philosophy supplied, they
entitled themselves to believe in countless inhabited worlds beyond
our own. Notice how in this case materialistic philosophical assumptions underwrite the aliens.

Aristotle (384–322 BC) and his followers tried to kill all this
extraterrestrial fun with Vulcan-like logic and alternative philosophical premises that allowed for only one necessary and eternal cosmos,
with just one inhabited globe at its center. They taught the existence of
four terrestrial elements: earth, water, air, and fire. The "element" earth
must fall down toward its natural place, the cosmic center. Humans
were the only rational animals in the universe. The vast but finite
heavenly realm was made of a special fifth element that was changeless
and therefore lifeless in a biological sense. Thus, in Aristotle's cosmos,
there was simply *no place* for ET. There was nothing—not even empty
space—beyond the immense sphere of fixed stars, and the immaterial
Prime Mover that contained and animated the whole cosmos.

Because Aristotelian physics seemed to explain more things, and
because it was somewhat easier to reconcile with Christianity than
atomism, most early Christian thinkers favored Aristotle's cosmology, along with similar perspectives that traced back to Plato. So late
ancient and early medieval Christian thinkers mostly opposed the
plurality of inhabited worlds, and in so doing allied Christianity with
one of the major ancient pagan cosmological traditions.

Given this background, it is unsurprising that around AD 260,
Pope Dionysius of Alexandria criticized the theory of Democritus and
Epicurus that everything is composed of atoms randomly arranged,
including the chance creation of "infinite worlds."[11] But note that he
did not declare the plurality of worlds itself heretical, perhaps because
(although we have no record of it) he understood that God is capable
of creating, and free to create, as few or as many habitable planets or
universes as he pleases.

Basil the Great (ca. 330–379), well educated at Constantinople
and Athens in the natural sciences and mathematics, argued precisely this about God's freedom and power, especially as a challenge

to the necessitarianism of Aristotle's cosmology. In his commentary on Genesis, which synthesized natural and revealed knowledge, Basil reasoned that God could create many universes, each with its particular comprehensible natural order.[12] But he did not assert that God actually did create other universes; there were no compelling reasons to draw that conclusion either. Basil, like many Christian thinkers since, recognized the *contingency* of our universe: states of affairs that *might have been otherwise* but that are still amenable to rational analysis. This, rather than the question of whether other universes actually exist, was the chief Christian focus.

Such contingent cosmic structure, grounded in God's omnipotent freedom, is *opposed to all* pagan worldviews that select other explanations for cosmic order and life, most notably chance (rare, unpredictable, unchosen events) and/or necessity (what always or regularly happens). Basil was quoted by many thinkers all the way up through the seventeenth century. This Christian intellectual tradition grounded the intelligibility of our universe (and others that might exist), a precondition of scientific investigation, in the supreme rationality, freedom, and love of the one Trinitarian God.

Alberto Martinez, a leading authority on the history of heresy, notes that after Pope Dionysius, "Patristic philosophers increasingly debated whether the Earth is the only world that exists." But then, after failing to mention the enormously influential Bishop Basil, he writes that "soon, some Christian theologians denounced the notion that there exist many worlds as heretical."[13] This is misleading. His first example is Philaster, Bishop of Brescia.[14] Among the 156 supposed "heresies" that Philaster attacked, besides the plurality of worlds, was the belief that stars have a fixed place in the sky rather than being set in position each evening by God.[15] Good luck finding more than just a few other ancient and medieval Christians who thought that *this* was heresy! Philaster was exceedingly loose with the charge of heresy, and for this and other reasons many scholars have identified him as incompetent.[16]

Let us now examine the more reputable characters in Martinez's heresy chronicle. In 402 Saint Jerome retrospectively labeled some of Origen's (ca. 184–254) teachings as "most heretical," including "worlds innumerable, succeeding one another eternally." Jerome noted that this teaching was associated with Origen's belief in the transmigration (reincarnation) of souls through successive worlds.[17] The Fifth Ecumenical Council (553) subsequently declared Origen a heretic on multiple grounds, but belief in a "plurality of worlds" was not among them. So this council did *not* affirm Jerome's "most heretical" label (or even just plain old "heretical") for Origen's plurality of worlds. The 553 council was probably aware of the fact that the transmigration of souls is closely related to the plurality of worlds. Jerome himself had made this connection explicit in 402. So the 553 heresy labeling of soul transmigration, but not of many worlds, is remarkable. The council had clear precedent in Jerome to declare *both* heretical in an interrelated matter, but it did not do so.

In 748 Pope Zacharias denounced as "perverse doctrine" the claim, which the abbot Virgilius allegedly made in Salzburg, that "there is another world & other men beneath the earth, another Sun & Moon."[18] Martinez takes this to include another *extraterrestrial* world with its own sun and moon. But John Carey argues convincingly for a more accurate translation in which the last part of the sentence is rendered "or even the sun and moon."[19] Carey explains that if, as this better translation suggests, we take the "sun and moon" to be our local luminaries that give light to both hemispheres of earth, then the entire passage makes better sense as referring to the controversial idea of the "antipodes"—the hemisphere opposite the known world, inhabited by human-like creatures not descended from Adam and Eve.[20] Pope Zacharias probably denounced this notion of antipodes, not extraterrestrial rational creatures, as "perverse doctrine."[21]

Many Worlds Outside the Aristotelian Box

From the thirteenth century onward, "other worlds" became a frequent topic of disputation in medieval universities. The leading late medieval thinker Thomas Aquinas (1224–1274), in his *Summa Theologica*, examined the possibility of more than one Aristotelian-type cosmos. Although he sided with Aristotle in rejecting the idea, he also insisted that the singularity of our world does not imply a limitation to God's creative omnipotence.

Some thought that Aquinas had not gone far enough to recognize God's omnipotence in the face of Aristotle's philosophy. In 1277 these critics worked with the bishop of Paris, Stephen Tempier, to condemn 219 propositions related to this topic and other important issues. One of those condemned propositions, #34, stated, "That the first cause [i.e., God] could not make several worlds."[22] Shortly after this condemnation, many scholars constructed arguments showing that God *could* create many worlds beyond our earth-centered system.[23] In effect, this Church-imposed condemnation created a lot of hypothetical real estate in which ET could live.[24] A large conceptual portal to other worlds had opened.

Nicole Oresme (1325–1382), a philosopher and Aristotle critic who became the bishop of Lisieux, conceived of a series of different worlds located in separate segments of time or space. He concluded: "God can and could in His omnipotence make another world besides this one or several like or unlike it. Nor will Aristotle or anyone else be able to prove completely the contrary. But, of course, there has never been nor will there be more than one corporeal world."[25]

Another world "like" ours would have intelligent life, but Oresme did not pursue this thought. Nicholas of Cusa (1401–1464) did.

In *Of Learned Ignorance* (1440), Nicholas concluded that the earth moves so as to give the illusion of the cosmos's spinning about us.[26] He further reasoned that our earthly home is not radically different from the rest of the universe, as Aristotle claimed. If earth is similar to the other celestial bodies—to a planet like Mars or even to the stars—why not also in regard to the presence of organisms, even alien "men" like or unlike us?[27]

So Nicholas believed in aliens! Surely the Catholic Church branded him a heretic. No, instead, eight years later, it made him a cardinal.[28]

Whereas Nicholas of Cusa did not address how Jesus relates to aliens, the highly regarded Franciscan scholar William Vorilong (d. 1464) dove eagerly into this pool of speculation:

> If it be inquired whether men exist on that [alien] world and whether they have sinned as Adam sinned, I answer no, for they would not exist in sin and did not spring from Adam.... As to the question whether Christ dying on this earth could redeem the inhabitants of another world, I answer that he is able to do this even if the worlds were infinite. But it would not be fitting for Him to go unto another world that he must die again.[29]

Neither was Vorilong declared a heretic.

So both ET lovers and haters existed within historical Christendom. Yes, there were more haters, but let us not forget the dazzling lovers up through 1450—with many more to follow in later periods.[30] During this period not a single pope or major church council declared ET or many worlds to be heretical.

Bruno's Souls Transmigrating Through Many Worlds

The case for Christianity's being against the plurality of worlds got a boost when Pope Gregory XIII, in the throes of the Counter-Reformation, commissioned an updated *Corpus of Canon Law* (1582). This was partly to better prosecute proliferating heresies. Consider this resulting passage: "There are also other heresies without author and without names," among which is "having the opinion of innumerable worlds."[31] Although the pope ordered this new edition to be prepared and used, he did not *write* it. We do not know if he read and specifically affirmed the statement that condemned innumerable worlds. For all we know, only a single church bureaucrat involved in the project felt passionate about including that particular sentence fragment in the massive three-volume work.[32]

In the decade after the revised canon law appeared and before his 1592 arrest in Venice, Bruno published at least nine books asserting the plurality of worlds.[33] In *On the Infinite Universe and Worlds* (1584), he cited Cardinal Nicholas of Cusa, clearly not a heretic, as an earlier proponent of many worlds.[34] During his trial Bruno cleverly evaded or equivocally disowned many heresy accusations, but he stood firm on his belief in many worlds. He insisted that he had always advocated the plurality of worlds "without indirectly meaning to reject the truth according to the faith."[35] Although he failed to convince his inquisitors of the acceptability of many worlds, there was sufficient historical precedent for making this attempt. Trained in prestigious Dominican institutions, Bruno probably knew at least some of the history of the broad range of views about the heretical status (or not) of the idea of many worlds.

Soon after Bruno's death, the prominent French scientist-theologian Marin Mersenne (1588–1648) analyzed his legacy. In dialogue with

Bruno's ideas (and with many other scientists through his frequent letters), Mersenne declared in 1623, echoing Saint Basil, that many worlds might exist owing to divine freedom. Consequently, he argued that a plurality of worlds was a legitimate, nonheretical topic for "*theologia argumentatrix*" (theological disputation).[36] He even noted, in line with my historical sketch, that no ecumenical council had ever declared many worlds heretical.[37]

Mersenne, however, strongly resisted Bruno's *necessitarian theological* argument for infinite worlds, noting its roots in ancient philosophers who had a "very confused notion of God and [were] full of errors."[38] Mersenne said such a view of divinity amounted to heresy because it denied God's freedom to create as many or as few worlds as he pleased. The only necessity with respect to God, he argued, was *internal* to God himself as Father, Son, and Holy Spirit.[39] Mersenne concluded that Bruno had gone "well beyond" the completely legitimate cosmological pursuit of "infinite worlds in the stars" and "attacked the Christian Truth."[40] He declared Bruno "an atheist, who has been burnt in Italy for his impieties."[41] So Mersenne saw Bruno not as a martyr for science but as a martyr for atheism.

Breaking from many other historians, Alberto Martinez shows that the issue of many worlds was an important component of Bruno's trial. But to what degree did the inquisitors deem this issue important not as "science" but because it was rooted in pagan philosophical theology, as Mersenne concluded? Martinez argues that many worlds might have been a sticking point in part because of its connection to religious belief in the transmigration of souls.[42] This suggests that Bruno was a martyr not for science but for pagan religion and pagan philosophy. As historian Maurice Finocchiaro puts it, Bruno's trial was much more "philosophy versus religion" than "science versus religion."[43]

According to Finocchiaro's essay, which Martinez cites approvingly, Bruno's heresies probably included the eight theses in the (now lost) official condemnation list "concerning which there was no question that Bruno did hold them, but only whether and when they had been formally declared to be heretical."[44] But his heresies included many other charges "regarding which the question was whether Bruno did really hold them," although they were clearly heretical.[45]

Bruno's Anti-Trinitarian Theology

Bruno admitted to doubting the Trinity, which stands at the center of Christian doctrine. He argued that his doubts took the form of internal private struggles, not public displays of disbelief.[46] But the Inquisition discovered that many of his books contained highly unorthodox views with potentially heretical implications. Those objectionable teachings included: "the stars are animate, that is, possess rational souls"; "the earth is animate, that is, possesses a rational soul"; and "the Holy Spirit may be identified with the soul of the universe" (here is a *public* statement that sounds pantheistic and undermines the Trinity).[47] Under interrogation, Bruno strained to justify his view of souls:

> Speaking as a Catholic, they [souls] do not pass from body to body, but go to Paradise, Purgatory or Hell. But I have reasoned deeply, and, speaking as a philosopher, since soul is not found without body and yet is not body, it may be in one body or another, and pass from body to body. This, if it be not true, seems at least verisimilar [likely true], according to the opinion of Pythagoras.[48]

Reasoning from his Pythagorean view of the soul, he believed that death was not to be feared but was merely a transition from one body to another. This offers further evidence that primarily divergent opinions in theology, not science, condemned Bruno.

At the September 9, 1599, meeting of the Inquisition, Pope Clement VIII declared that Bruno must retract his admitted heresies and that the trial documents should be reexamined for additional heresies. When Bruno heard this news the following day, he said he was ready to confess his previously interrogated opinions as errors, but he also submitted a memorandum addressed to the pope in which he defended himself regarding those very opinions. This bipolar maneuver triggered a cascade of events that led to his execution.

Finocchiaro argues that as the trial drew to a close, Bruno "could not bring himself to abjure philosophical opinions which he felt were not heretical and had never been formally declared to be heresies."[49] Bruno had at least this much right: within Christendom there existed a range of opinions about whether the plurality of worlds was heretical.

The same degree of historical latitude did not exist for most of Bruno's other putative heresies. For example, sometime in September 1599 the inquisitors discovered that Bruno's book *The Expulsion of the Triumphant Beast* (1584) contained "an indictment of Christianity and a thinly disguised mockery of Jesus" (in the words of scholars William R. Shea and Mariano Artigas).[50] Mocking Jesus in sixteenth-century Europe was not a good idea.

Ingrid Rowland, in her biography of Bruno, estimates that even in the earlier Venetian phase of Bruno's trial (before his transfer to Rome), "it must have been clear to these Christian inquisitors that Bruno was no longer defending a Christian position" when he explained his doubts about using the term "person" in regard to Jesus as the second person of the Trinity:

> To get to the individual point about which I have been asked concerning the divine persons, that wisdom and that son of the mind whom the philosophers call the intellect and the theologians the Word, if one ought to believe that this took on human flesh, then, philosophically speaking, I have never understood this, but have doubted it and held the belief with inconstant faith.[51]

Although Bruno believed in most of Christ's ethical teachings, he deemed unintelligible the Christian claim that Jesus was unique in being both fully human and fully divine. All souls, even animal souls and planetary souls, were equally instantiations of the divine world soul, he thought, and so only in this deflating (and clearly heretical) sense could he affirm the divinity of Jesus. Primal stupidity, not primal sin, was our problem, he felt.[52] Consequently philosophy, not Jesus, was the solution—or so he hoped. No wonder so much trial time was devoted to this issue as Bruno found ever more ways to evade questions and to appear, often, to hold a Christian view of Jesus.[53] But in truth, Bruno thought Jesus was just one more soul on an infinite transmigrating journey through an infinite number of bodies and worlds.

Apparently Bruno bet his soul on this infinite journey. According to Gaspar Schoppe's eyewitness testimony, the burning Bruno was presented with a crucifix, but "he turned his face away to reject it."[54]

Did Bruno Acquire Correct Beliefs About the Cosmos by Rigorous Investigation?

Although Bruno got many things wrong (as we all do), he also was the first to articulate a number of surprisingly correct beliefs about the cosmos. But does getting the right answer about nature, in regard to a few isolated topics, make him a model scientist or scholar?[55] We must identify by what standards he justified those beliefs.

In Bruno's clever literary dialogue *On the Infinite Universe and Worlds*, he has one of his characters affirm: "There are then innumerable suns, and an infinite number of earths revolve around those suns, just as the seven we can observe revolve around this sun which is so close to us."[56] Today it is beyond reasonable doubt that countless stars in the sky are actually suns, like our host star, around which planets orbit. So Bruno got this right! Because Bruno's "earths" included our moon, he scored seven planets from Mercury (closest to the sun) to Saturn (the outermost planet known at the time).

In Bruno's day, however, it was impossible to know confidently whether the earth revolved around the sun, or the opposite. The definitive evidence needed to settle this question was not yet available. It also was impossible to know whether planets orbited other stars. We now know of thousands of such extrasolar planets, and the number of additionally detected planets grows rapidly. Definitive scientific technology for such detections did not emerge until the 1990s. But in Bruno's day the case for extrasolar planets was extremely weak, propped up with numerous questionable assumptions. So the way that Bruno acquired and defended his belief in extrasolar planets fell short of reputable science by the standards of both the sixteenth century and today.

Champions of the "Bruno as martyr for science" meme also generally avoid discussing the animistic aspects of his writings.[57] Bruno stressed the need to consider physical causes, which in his day, as a holdover especially from Aristotle, included analysis of the actions of souls. His own attempt at this was so deeply entangled with his animistic view of nature that very little of his cosmology actually contributed to the growth of science. His explanations often invoked the world soul and individual souls in celestial objects. For example, Bruno claimed that the Copernican motions of earth resulted largely because the earth's rational soul sought to sustain itself in a stable condition.[58]

Although Bruno accepted a vaguely Copernican cosmos (earth moves around the sun), he largely disdained the very mathematical analysis of nature that made the Copernican achievement most valuable.[59] Here is what Bruno professed about Copernicus and his mathematical astronomy:

> But for all that he did not move too much beyond them [his predecessors]; being more intent on the study of mathematics than of nature, he was not able to go deep enough and penetrate beyond the point of removing from the way the stumps of inconvenient and vain principles, so as to resolve completely the difficult objections, and to free both himself and others from so many vain investigations, and to set attention on things constant and certain.[60]

Bruno sometimes presented himself as building on a Copernican foundation, but he attributed to Copernicus (and personally adopted) an astronomical belief that was embarrassingly at odds with Copernicus's own book and with the observational evidence available in his day and ours.[61]

Bruno constructed an alternative post-Copernican cosmology that eventually belly-flopped as scientists went elsewhere to unveil nature's secrets.[62] Bruno's short-lived cosmology included a recklessly imprecise thermodynamic account of planetary motion that demanded just two categories of celestial objects: hot suns and cold non-suns that move around a hot sun for both animistic (soulish) and mechanistic (hot-cold thermodynamic) reasons. His cosmology offered no place for cold moons orbiting cold planets. Nor could his theory account for the phases of the moon. Bruno had opened the door to a higher cosmology that could not be bothered by the tedious details of "the mathematicians," as he dismissively called the astronomers, both the countless geocentric and very few heliocentric astronomers of his day.

To be sure, "the mathematicians" did not take kindly to the presumptive superiority of this man from the town of Nola near Naples—"the Nolan," as he frequently called himself. The greatest of these mathematical astronomers, the Danish Tycho Brahe, scribbled a cruel pun on the flyleaf of his copy of Bruno's *Camoeracensis Acrotismus* (The Pleasure of Dispute, 1588): "Nullanus, nullus et nihil" (The Nullan [the Nolan], the Nobody, the Nothing). To be clear he added, "Names often agree well with their objects."[63]

Bruno was on the fringes of respectable astronomical company. Small wonder that Oxford scholars ran the boastful migratory philosopher out of town after his lectures there. In fact, just about everywhere Bruno went he quickly wore out his welcome. His heretical beliefs provoked his serial excommunication by Catholics, Calvinists, and Lutherans (in that order). But some of his migratory behavior came from Bruno's eruptive habit of mocking almost anyone with whom he disagreed.

More Bruno Confusion

So what is the basis for the idea of Bruno as a martyr for modern science? It is rooted in a superficial and distorted image of Bruno.[64] Some scholars have creatively constructed a Bruno who displayed "a talent for prophetic intuition" and urged us to call him a "forerunner" of today's science, including even quantum physics and its Heisenberg uncertainty principle.[65] But there is no good reason for calling Bruno a hero of modern science as a reliable way of knowing the world. Although Bruno might stand legitimately as a tragic hero for free speech, his cosmic speculation contributed very little to the long-term growth of science.

What about the basis for Bruno's belief in an infinite universe without beginning, which was a dominant theme in several of his books? He primarily worked out his position on this through an exercise in philosophical theology. That discipline *can* be practiced rigorously, but even if he practiced it well, this would not make him the sort of investigator that most champions of Bruno as "forward-looking scientist" have in mind.

In any case, Bruno did not practice the discipline rigorously. His long road to cosmic infinity began with his anti-Trinitarian theology. Amplifying and modifying ancient pagan philosophical theology, Bruno claimed that God's power enables and *requires* him to create worlds endlessly.[66] Everything that God can create, he must actually create; he has no choice. But Bruno did not satisfactorily specify the logical necessity that allegedly bound God to create compulsively and endlessly.

The difference between Bruno's cosmic infinity and that of Nicholas of Cusa is striking. For Nicholas, emulating historic Christian theology, God incarnate in Jesus is the only embodied full expression of God's divinity, including God's power and so much more. Nicholas believed that even an infinite cosmos, such as the one he proposed, falls woefully short of the fullness of God in Christ. Bruno reversed this understanding: the infinite cosmos itself is the only full embodied expression of deity and amounts to something like the body of God. Jesus is just one more local embodiment of the infinite divine world soul.

This is pantheism, not Christianity. The cosmos, as the body of God, is the mediator between the inaccessible face of God and humanity. Bruno denied Jesus that role. No wonder he mocked the historic Jesus, and eventually was identified as an unrepentant heretic. Burning Bruno was a cruel and terrible idea, but a heretic he was.

Bruno's Legacy

Bruno rejected the historic Jesus, the Trinity, as well as other key components of Christian theology. His infamous trial before the Inquisition found him unrepentantly guilty of at least some of these heresies. The trial only tangentially and briefly addressed his Copernicanism. But his infinite cosmos of many inhabited worlds became a major conversation piece, in large part because of its connection to his belief in the transmigration of souls and his substitution of an infinite cosmos for what had been reserved in Trinitarian theology for Jesus as God incarnate and redeemer of repentant human souls.

Bruno was not a martyr for science. He died chiefly for a series of conventional heresies as well as for a few cleverly reconfigured heresies—some entangled with infinite-worlds ideology that came out of his naturalistic philosophical theology. His habit of mocking and confusing opponents also did not help him avoid the fires of Rome.

His true story is heartbreaking nonetheless. Perhaps an apology from the Roman Catholic Church, informed by the latest historical scholarship, is forthcoming. Will there also be an apology from those who continue to parade him as a martyr for modern science? A sober historical assessment of our ancestors—giving both recognition and critical judgment where it is due—is the civil, right, and loving way to live. May Bruno's legacy remind us of that truth.

5

GAGGING GALILEO

[The Catholic Church had been] torturing scholars to the point
of madness for merely speculating about the nature of the stars.
—*Sam Harris*, The End of Faith: Religion, Terror, and the
Future of Reason *(2004), 105*

The Christian Church fought bitterly throughout its history—
and is still fighting today—to impede scientific progress. Galileo, remember, was nearly put to death by the Church for constructing his telescope and discovering the moons of Jupiter.
—*David Mills*, Atheist Universe: The Thinking Person's
Answer to Christian Fundamentalism *(2006), 48*

The two epigraphs are completely false. These authors are leaders
in the New Atheism movement, which is committed to the idea
of total warfare between science and religion, regardless of the historical evidence.[1] According to the warfare thesis, which long predates
the New Atheism, Galileo Galilei (1564–1642) represents unbiased
scientific objectivity and the Catholic Church stands for ignorant
superstition that has hindered the growth of science. The inaccurate
account of Galileo's being imprisoned and tortured by the Catholic Church originated and persisted mainly owing to the manner
in which the Catholic Church announced the outcome of Galileo's
trial, and also because the Church did not allow public access to most

of the primary documents until the late nineteenth century. More recently, further exaggerations have developed to fit the requirements of the warfare narrative. According to this evolving storyline, Catholic officials rejected Galileo's telescopic observations and his rational arguments that had allegedly proved the Copernican system.

The real Galileo story is much more complex than this. But one thing is clear: the episode does not support the inevitability of a deep conflict between science and Christianity.

Galileo's Startling Telescopic Discoveries

By about December 1, 1609, Galileo had constructed a telescope that magnified objects twenty times. This is the strength of a decent pair of binoculars today. His telescope also had a narrow field of view: when he pointed it at a full moon, he could see a little less than 25 percent of the moon's illuminated surface in one glance.[2] Still, his telescope marked a major improvement over the original device, invented in Holland. Galileo could see that the moon seemed to have moun-

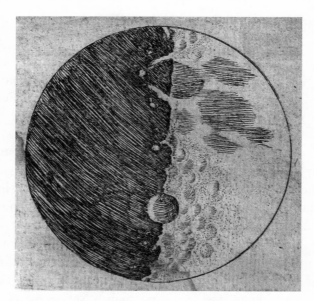

Figure 5.1: A lunar drawing from Galileo's Starry Messenger *(1610). (Photograph courtesy of the History of Science Collections of the University of Oklahoma)*

tains and valleys—many with circular shapes now known as craters (Figure 5.1). Although he overestimated lunar mountains at almost five miles tall, he left humanity little doubt that the moon was not a perfectly smooth sphere, as Aristotle had taught. "I thank God from the bottom of my heart that he has pleased to make me the sole initial observer of so many astounding things, concealed for all these ages," he wrote the Tuscan secretary of state on January 30, 1610.[3] Piety and diplomacy kissed.

In July 1610 the Grand Duke of Tuscany (an Italian region just north of the Papal States) appointed Galileo his chief mathematician and philosopher. The appointment was for life. This honor allowed him to return to his country of birth, but it came only after strategic acts of kindness toward the ruling Medici family. Galileo gave the name "Medicean Stars" to the four satellites (later called moons) that he discovered revolving around Jupiter. He first thought they were ordinary stars, but then he observed them change positions relative to Jupiter. His telescope also allowed him to see that the number of fixed (ordinary) stars vastly exceeded naked-eye estimates.

In Tuscany, Galileo observed that the planet Venus goes through phases analogous to those of the moon: from virtually full, to half, to crescent, and back to half and virtually full. The phases of Venus could be explained if the planet revolved around the sun. So this finding refuted the ancient geocentric Ptolemaic system, which had Venus going around a motionless earth, not around the sun.

But the discovery did not prove the Copernican system. Although Copernicus had argued that Venus revolves around the sun (in a path closer to the sun than earth's annual revolution), Tycho Brahe had proposed a third system that also proved consistent with Galileo's telescopic wonders. In the Tychonic system (Figure 5.2, next page), the planets revolve around the sun, which in turn revolves around the earth. When Galileo showed that Jupiter revolves around the sun while also carrying its four moons for the ride, this supported the Tychonic system's contention that the sun could revolve around a central point while carrying with it other celestial objects (planets) that revolve around it. Galileo had emphasized only the support the moons of Jupiter lent to the Copernican system, as that finding showed it was reasonable to believe that maybe earth was not alone as a planet that carried with it a revolving moon.

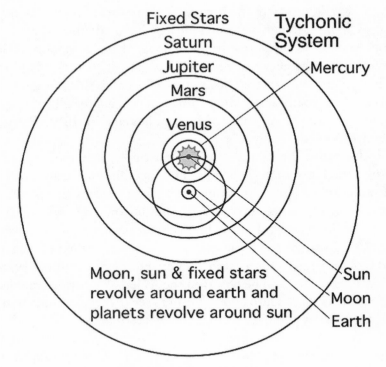

Figure 5.2: The Tychonic system

The astronomers at the Roman College, the flagship Jesuit educational institution, embraced the Tychonic geoheliocentric system. The Tychonic system fit common sense: earth seems to be at rest rather than moving thousands of miles per hour. Furthermore, a central stationary earth fit the prevailing Aristotelian scientific theory of how ordinary material objects with weight behave. Heavy (earthy and moist) things fall downward, toward the cosmic center, naturally making the roughly spherical object on which we live: earth and its surface water. No wonder that after the marvels of the telescope, most university scientists embraced Tychonic cosmology.[4]

There also appeared to be theological advantages to a geocentric cosmology—Tychonic or Ptolemaic. At least this is how the vast majority of Catholic and Protestant leaders assessed the situation up through Galileo's lifetime. The Bible seemed to affirm a stationary earth around which the sun moves. Although opposition to Galileo first came from Aristotelian professors, pastors and theologians did not welcome having to reinterpret the Bible in light of new scientific ideas.

Exchanging Letters: Galileo and the Grand Inquisitor

Even if church officials typically embraced geocentric cosmology, that did not mean they reviled Galileo. On the contrary, many Catholic leaders respected him. Roman intellectuals with close Vatican ties, including the leading Jesuit astronomer, Christoph Clavius (1537–1612), had helped a young Galileo land his first two university positions—Pisa in Tuscany, then Padua in the Republic of Venice. Galileo returned to Rome in 1611 to celebrate his telescopic discoveries with Pope Paul V, many cardinals, and others. Numerous banquets were held in his honor. The Roman College even granted him the equivalent of a modern honorary doctorate in a lavish ceremony.[5]

The respect among Catholic leaders continued through Galileo's trial and final days, though the number of his supporters dwindled. The reasons for this reduction in support are complicated. The pride and vanity of both Galileo and Pope Urban VIII played contributing factors.

Galileo did have to answer theological objections to Copernicanism. In response, he wrote two widely circulated letters. The first, in 1613, went to his closest disciple and friend, Benedetto Castelli, a Benedictine monk and professor of mathematics at Pisa. Galileo enlarged that work in a 1615 letter to the Tuscan Grand Duchess Christina. In the Christina letter he pondered the principles of biblical interpretation specified by the Council of Trent (1545–1563). That council was the Catholic response to the Protestant Reformation. The interpretive principle especially at issue declared that the Bible should be understood in a way consistent with the consensus of earlier church theologians—the Holy Fathers. Galileo, however, argued that this principle applied only to cases in which the Holy Fathers explicitly examined a specific question with appropriate intellectual tools. Such was not the case regarding whether the sun revolved around a stationary earth. Most previous commentators merely assumed that the Bible reflected a commonsense understanding of a motionless earth. They did not adequately address this question, and so their less-than-rigorous consensus about the sun's revolving around earth did not present a binding biblical interpretation, he insisted.[6]

Galileo also noted the long and respected theological tradition of understanding biblical descriptions of natural phenomena as reflecting how things appear to the human observer. The Holy Spirit, inspiring the human biblical authors, did not intend to teach cosmology or the subtle mechanics of nature. Such things were unknown to the original audience and would have only confused them if the Bible had revealed some of these things.[7] Galileo quoted Cardinal Cesare Baronio (1538–1607) to make the lesson memorable: "The intention of the Holy Spirit is to teach us how one goes to heaven, not how the heaven goes."[8] This implied no error on the part of the Bible. Galileo pointed out how Copernicus himself had taken this approach in his book *On the Revolutions of the Heavenly Spheres* (1543). He used common expressions like sunrise and sunset, even though he argued that such appearances actually resulted from the earth's rotating rather than from the sun's revolving.[9]

As to cosmology, Galileo argued that the Bible was not intended to teach *any* of the candidates for world systems in the seventeenth century: Ptolemaic, Tychonic, or Copernican.[10] God expected us to resolve cosmological questions by using the rational tools he gifted to us. Observational tools like the telescope were also required. Galileo put down his pen hoping that his oversized letter would convert the luminaries of his own beloved Catholic Church.

Between the time of Galileo's letter to Castelli and the letter to Christina, Cardinal Robert Bellarmine (1542–1621) wrote a letter to the Copernican enthusiast and theologian Paolo Antonio Foscarini. It was intended as a response to Galileo's letter to Castelli. Thus Galileo used his letter to Christina also to reply to Bellarmine's letter to Foscarini.

It is important to understand the theologian with whom Galileo indirectly dialogued. In 1576 Bellarmine began his rise in Rome as the chief Jesuit defender of Catholicism against the Protestants to the north. He became a cardinal in 1599 and served a prominent role in the last phase of Giordano Bruno's trial, including as one of the judges. Soon after 1600 he emerged as the leading Catholic theologian and inquisitor. For Protestants he was theological enemy number one. Yet he was also widely known for kindness—even to Galileo.[11]

Historians generally agree that he was very intelligent and a man of principle. The principles that burnt Bruno and gagged Galileo were faulty, no doubt. But even smart people known for kindness, whether ethical atheists or misguided theologians, can feel duty bound to do horrible things.

Scholars still debate how to interpret Bellarmine's April 1615 letter. In it, Bellarmine instructed Foscarini (and Galileo) that if a "true demonstration" were to firmly establish Copernicanism, then "one would have to proceed with great care in explaining the Scriptures that appear contrary."[12] He said he had "very great doubts" that such a demonstration of heliocentrism would ever be accomplished. "In the case of doubt one must not abandon the Holy Scripture as interpreted by the Holy Fathers"—that is, the Council of Trent's rule about the consensus of earlier church theologians.[13]

But what if scientific discovery were to remove reasonable doubt about the Copernican system? In this case, it appears Bellarmine said that we should go with Copernicus, and theologians would need to reinterpret the Bible accordingly. The earlier contrary consensus of the Holy Fathers would not be binding in this scenario.

Contrary to what some historians of science have concluded, Bellarmine's letter does not appear to oppose scientific progress in principle.[14] The cardinal was also correct in thinking that the Copernican system had not yet (in 1615) been proved beyond reasonable doubt. Even in 1633, when Galileo was put on trial, most scientists questioned sun-centered astronomy. Only much later did the Copernican system advance to a position beyond reasonable doubt.[15]

The 1616 Condemnation of Copernicanism

In February 1615 the Florentine Dominican friar Niccolò Lorini filed the first formal complaint about Galileo's Copernican advocacy with the Roman Inquisition. Then in March the Dominican Tommaso Caccini submitted a personal deposition with the Inquisition. This triggered an investigation that lasted about a year.[16] In December 1615 Galileo traveled the two hundred miles from Florence to Rome to defend himself, even though he had not been summoned. Although many influential Church officials received him well, the

Inquisition soon gave Galileo a private warning to abandon Copernicanism. Bellarmine delivered the personal message and later reported
that Galileo had promised to obey.

In March 1616 the Congregation of the Index, the office responsible for book censorship, issued a public decree declaring the idea
of a moving earth false, "altogether contrary to Holy Scripture,"
and a source of illegitimate "prejudice" against Catholicism. The
decree "completely condemned and prohibited" Foscarini's theological defense of Copernicus, to which Bellarmine had responded in
his April 1615 letter.[17] The Congregation of the Index temporarily
banned Copernicus's famous 1543 book, pending corrections. The
decree did not mention Galileo at all.

Why did the Church condemn Copernicanism? Did the decree
reveal a war between science and religion?

A May 2, 1633, entry in the diary of Galileo's friend Gianfran-
cesco Buonamici gives an important detail about the 1616 Index
meeting that crafted the Copernican condemnation. Two of the par-
ticipating cardinals successfully argued for a weaker censure than the
"heretical" label that the Inquisition qualifiers had recommended. So
the Congregation of the Index chose the milder phrase "altogether
contrary to Holy Scripture" for the public decree.[18] One of these cardinals was Maffeo Barberini, who was to become Pope Urban VIII
during the latter part of Galileo's life. In 1630, Pope Urban VIII
would tell his adviser Tommaso Campanella that "it was never our
intention [to prohibit Copernicus]; and if it had been left to us, that
decree would not have been made."[19]

Why, then, did Pope Urban VIII take a hard line on Galileo's
1632 book *Dialogue Concerning the Two Chief World Systems*?

Cardinal Barberini became Pope Urban VIII in 1623, two years
after Bellarmine's death. Galileo met with his well-educated friend
the new pope to see whether Urban would support his plan to further investigate Copernican astronomy. After six exploratory conversations over six weeks, Galileo detected sufficient freedom to begin
writing a book that would defend Copernicanism *implicitly* through
a dialogue among three fictional characters. The book did not present the pro-Copernican arguments as conclusive, and it voiced the
arguments against earth's motion, though they seemed much weaker.
As such, the *Dialogue* could appear to meet the Catholic Church's

demand, delivered through Bellarmine, that Galileo not "hold" or "defend" the theory of earth's motion.

But it soon became clear to the majority of Inquisition decision makers that Galileo had violated at least the spirit of Bellarmine's warning and the Index's decree. For example, Galileo vastly overstated the value of his leading argument for a moving earth: his theory of tides. He argued that water sloshed around within a container as the container moved and that earth was a big, and moving, container for the oceans. His theory specified one high tide and one low tide per day. Shortly before he finished his book, however, Galileo's theory collided with the inconvenient fact that each day brings two high tides and low tides. Galileo bandaged up his bleeding theory and hoped for the best. Perhaps the odd shapes and the varying depth of the ocean floor could account for the gap between his initial theory and what sailors reported.

Inquisition officials were concerned not just with Galileo's unbalanced treatment of the competing views of cosmology and physics. They also discovered in the file of the 1616 proceedings a special injunction that had prohibited Galileo from even *discussing* the earth's motion in *any* manner.[20] The *Dialogue* clearly violated that requirement. So he was summoned to Rome for trial.

The 1633 Trial of Galileo

The early 1630s were politically perilous for Urban VIII. Europe was in the middle of the Thirty Years' War (1618–1648), which had begun along Catholic-Protestant fault lines. The pope was especially troubled over the Catholic monarchs of France and Spain, who competed for control of the shrinking Holy Roman Empire. Ecclesiastical politics within the Vatican were also burdensome, driving the pope to questionable legal actions in some instances.[21] This was not a convenient time to wrangle over the edgy cosmology of a troublesome genius.

The Galileo affair became very personal for Urban VIII. Galileo had put the pope's favorite argument for doubting Copernicanism in the mouth of the *Dialogue*'s character Simplicio, which sounds like simpleton in Italian. Simplicio is persistently ill informed and less than politely reasonable. The pope's old friend had betrayed him. Urban VIII was not the only one to see it this way. Regional and local politics, Galileo's insensitivity, the pope's overreaction, and other peculiar factors flung the trial into orbit. Nobody enjoyed the turbulent ride.

Galileo did not spend months or years in jail, as is often claimed. When he arrived in February 1633 he enjoyed pleasant lodging and fine cuisine at the large residence of the Tuscan embassy. Roman officials had spared him the usual procedure of waiting for trial in the Inquisition prison. They did, however, order him not to socialize and to stay off the streets.

In April, when the trial began, he stayed in the apartment that the Vatican notary had vacated for him. The chef of the Tuscan embassy delivered meals to him.[22] He might have stayed in a prison cell for just a few days around the time of his condemnation. Even if true, this is nothing close to the exaggerated claims of imprisonment associated with the Galileo myth. Immediately after his condemnation he lived temporarily at several palatial residences before being allowed to return to Florence for the remainder of his life under house arrest. House arrest, although humiliating, meant that Galileo lived at his own comfortable country residence overlooking the city of Florence. It came with certain unpleasant social and intellectual restrictions that amounted to a partial gagging of Galileo. But, in spite of this, Galileo accomplished his most important theoretical scientific work during this last phase of his life. His book *Discourse on Two New Sciences* (1638) includes major new insights into the science of motion.

At the first hearing in April 1633, Galileo admitted that Bellarmine had warned him not to hold or defend Copernicanism. He denied receiving the newly discovered special injunction prohibiting him from even discussing the topic.[23] In his defense he handed over a signed certificate Bellarmine had granted him in 1616 that prohibited him only from holding or defending the theory of a moving earth. Galileo claimed that the *Dialogue* did not defend the earth's motion but rather surveyed the arguments for and against it. Later in the trial he even claimed that his book was aimed at refuting the Copernican theory.[24] That was obviously a lie.

As Maurice Finocchiaro, a leading authority on Galileo, observes, Bellarmine's certificate and certain legal irregularities with the special injunction led the Inquisition to offer Galileo a plea bargain: "they promised not to press the most serious charge (violation of the special injunction) if Galileo would plead guilty to a lesser charge (transgression of the warning not to defend Copernicanism)."[25] Galileo agreed and eventually admitted that his book gave readers the impression of a Copernican defense. But he denied that proselytizing for Copernicanism had been his intention. Error and conceit were to blame, he desperately claimed.[26] Error is unintended, but what about conceit?

The trial verdict declared Galileo guilty not of heresy but of a lesser offense, "vehement suspicion of heresy."[27] He had to retract his Copernican beliefs by reciting a humiliating statement prepared for him. The *Dialogue* was also banned. End of story?

Actually, that was just the beginning of centuries of Galileo stories. The Galileo affair has been reassessed through many subsequent generations in an effort to grasp the significance of this troubling and confusing event. It has been commonly misconceived as a typical expression of the inevitable warfare between science and Christianity.

The Legend of Galileo in Astronomy Textbooks

The Inquisition publicized the sentencing document and Galileo's forced confession as a warning to all. These documents implied, but did not explicitly state, that Galileo had been tortured and would spend more time in prison. This false impression lingered long. Public access to the documents needed to debunk the jail myth came about 150 years later (1774–75). The records needed to undo the torture story took about 250 years to enter the public domain (publications of the trial proceedings appeared in various editions from 1867 to 1878).[28] Given the documents available today, we are even more confident that Galileo's interrogation came with the threat of torture, but not actual torture. It is even fairly clear that, given his age and popularity, his accusers never actually planned to torture him.[29]

Maurice Finocchiaro's book *Retrying Galileo, 1633–1992* is the definitive study that traces three and a half centuries of reevaluating or "retrying" Galileo.[30] I will highlight a few turning points in how astronomy textbooks have retold and refashioned the Galileo story. This will complement Finocchiaro's account.

In chapter 2 we met the Presbyterian minister and University of Pennsylvania professor-provost John Ewing. He advanced the Dark Ages myth in one of the earliest English-language astronomy text-books (1809) to contain *any* of the six historic warfare myths in our survey. Ewing claimed that after Galileo's telescopic discoveries, Catholic priests perpetuated the anti-intellectual mentality of the Dark Ages by forcing Galileo under torture to "deny the truth of what he and many thousands had seen with their own eyes."[31] Actually, there was considerable consensus on what was telescopically seen but much less agreement on what it meant.

Ewing did not claim Galileo spent much time in prison; Inquisition documents released about thirty-five years earlier had made that claim untenable. But the torture thesis was still believable until about 1870, and it made its way into this textbook. So the first astronomy textbook in my sample to contain part of the Galileo myth fits into Finocchiaro's history as a moderately responsible report of the Galileo affair given what was known publicly at the time.

Still, Ewing failed to mention how the Jesuits of the Roman College celebrated Galileo's telescopic discoveries. He also failed to grasp how none of this new knowledge, including the moon-like phases of Venus, conflicted with the geoheliocentric Tychonic system that the Jesuits championed. In this regard Ewing was guilty of the chronological snobbery and pseudo-history that C. S. Lewis later showed comes from avoiding primary sources—"old books."[32] Many astronomy textbooks since have treated the Galileo affair inaccurately.

Soon after Ewing, the British textbook author William Phillips erroneously claimed that Galileo was "imprisoned" for "asserting the truth" of Copernicanism. He added the jab "But what may not be expected from superstition and bigotry!"[33] Even though the jail thesis had become untenable by then, Phillips perpetuated it anyway.

In 1831 the Episcopalian minister-educator John Lauris Blake presented Galileo as an intellectual giant who fought widespread ignorance reminiscent of the supposed Dark Ages:

> Galileo was the victim of the persecuting spirit, which prevailed at that age of ignorance and superstition. When passed the age of three score and ten years, he was obliged by the priests, standing upon his knees, over the Bible, to disclaim belief in a sys-

tem to which he had devoted his days.... Subsequently to this, merely because he had the honesty to maintain that the Earth turned on its axis, he was condemned by a board of Cardinals to perpetual imprisonment. He did not long survive the loss of his liberty. He died at the age of eighty-four years.[34]

Note the erroneous jail thesis and the false impression that Galileo was persecuted merely for his scientific "honesty." Many honest and intelligent scientists advocated the alternative Tychonic system with arguments at least as respectable as those of Galileo.

Galileo was the principal hero in George G. Carey's *Astronomy* (1824), introduced as an early myth-laden textbook in chapter 1. Note the vastly overrated assessment of the arguments that Galileo launched, despite the fact that the Tychonic system was a strong competitor: "By the observations and reasoning of Galileo, the system of Copernicus acquired a probability almost equivalent to demonstration. By espousing the opinions of Copernicus, he drew on him the vengeance of the Inquisition, who decreed that he should pass his days in a dungeon; but he was liberated after the expiration of a year."[35]

Notice also that we have descended from a prison cell to a dungeon. The legend has grown.

In the second half of the nineteenth century, one of America's leading astronomers, Ormsby Mitchel (1809–1862), published the textbook *Popular Astronomy*. In it he described the alleged intellectual landscape of Europe just after Galileo's telescopic discoveries in 1610: "The scientific world was just in a transition state. The most honest, intelligent, and powerful minds had already adopted the Copernican theory, but in the universities and other schools of science, as well as in the church, the system of Ptolemy still reckoned among its supporters a host of learned and dignified men."[36] Mitchel, following a familiar pattern, mistakenly suggested that the Copernican system was virtually proved by this time.

This same misconception marks Edmund Beckett Denison (1816–1905) and Pliny Chase's (1820–1886) *Astronomy Without Mathematics* (1871). Referring to the same transitional moment, they asserted, "Then came Galileo...who...invented the telescope...and he found other proofs of Copernicus's theory, and (as is well known) was imprisoned for three years by the Roman Inquisition for publishing them."[37]

Although Galileo improved the telescope, he did not invent it. He often outperformed others at interpreting what could be seen through the telescope, such as in his detection of Jupiter's moons.

Joel Dorman Steele (1836–1886), one of the most prolific American textbook authors of the nineteenth century, intensified the Galileo myth. His frequently reprinted astronomy textbook stated: "Many refused to look through the telescope lest they might become victims of the philosopher's magic. Some prated of the wickedness of digging out valleys in the fair face of the moon. Others doggedly clung to the theory they had held from their youth."[38]

It is true that Galileo complained that some famous natural philosophy (science) professors at the University of Padua had refused to look through a telescope.[39] This behavior, though, probably resulted from a closed-minded commitment to Aristotelian physics rather than any theological view. One of the telescope-refusing scientists at Padua that Galileo mentioned was Cesare Cremonini (1550–1631), whom J. L. Heilbron in a biography of Galileo identifies as a "popular professor of philosophy at Padua, friend of Galileo, constantly in trouble with the Inquisition for his faithful teaching of Aristotle."[40] So both Galileo and the Inquisition criticized his excessive Aristotelian views.

Furthermore, there is no record of priests or theologians refusing to look through a telescope. In fact, when Cardinal Bellarmine asked about Galileo's telescopic discoveries, the Jesuit astronomers at the Roman College *confirmed* their accuracy. Bellarmine had himself gazed at the heavens through a telescope, probably because he wanted to be informed about discoveries of relevance to theology. The only telescopic interpretation of Galileo that the Jesuit astronomer Christoph Clavius contested was the claim that our moon's surface is irregular, with mountains and valleys. Several of his astronomy colleagues at the Roman College agreed with Galileo's interpretation.[41]

Even before Galileo's telescopic discoveries, Bellarmine had already concluded that heavenly bodies shared some of the imperfect features of our humble terrestrial realm. He based this opinion on the Bible. For example, Psalm 102:25–26 says both heaven and earth "wear out like a garment." Kepler invoked this passage to show that science and Scripture had converged on the same answer: the stars are not imperishable, as Aristotle had thought.[42] This seems to be a rare exception to the general interpretive rule regarding the Bible's accommodation to *appearances only* when treating natural phenomena. Both Kepler and Galileo had passionately defended such a general rule.

Many nineteenth-century astronomy textbooks perpetuated exaggerated versions of the Galileo affair. This trend continued until the past few decades, during which, I am happy to report, this particular myth has notably retreated in textbooks. Regardless, the New Atheism movement has helped revive the Galileo myth. It lives on in popular literature.

Galileo's Legacy for Science and Christianity

In the end, the Galileo affair, although embarrassing for the Catholic Church, does not support the common belief that Christianity typically suppresses the growth of science. Similar to his fellow Christian contemporary Kepler, Galileo was guided in his scientific work by the belief that God composed his book of the cosmos "in mathematical language." God's cosmic book "is constantly open before our eyes," Galileo assured his readers in his book *The Assayer* (1623).[43] Those readers included the learned and appreciative Pope Urban VIII, to whom Galileo dedicated the book.

The Galileo affair was not a simple or inevitable instance of science versus Christianity. Many complicated alliances and personal idiosyncrasies came into play. Contingency, not historical inevitability, was at work. The majority of Church leaders had allied themselves with the majority Aristotelian scientific viewpoint of the day. Together they opposed Copernican astronomy, which a theological and scientific minority held. If Galileo had been more tactful, modest, and patient in his attempt to reform his own church, there might have been no trial of 1633. Minority scientists such as Galileo argued that a heliocentric cosmos was scientifically superior. But given the scientific data available up through 1633, the Copernican system had *not yet* been shown to be superior to the Tychonic system of astronomy. Tycho Brahe's theory included many of the most defensible parts of the other two theories, and was endorsed by the Jesuit astronomers in Rome. Galileo strategically sidelined the Tychonic system in his *Dialogue Concerning the Two Chief World Systems*.[44] He was a master rhetorician. He made the arguments for the Copernican system seem stronger than they were at the time. This is part of what got him in trouble with the Inquisition.

Although Galileo said that God's "book" of creation "is constantly open before our eyes," accurately reading that book can be challenging. Skilled interpretation required telescopic wisdom and the newly developed techniques of mathematical physics, Galileo and Kepler argued. God's other book, the Bible, used ordinary observational expressions "of appearance" when referring to the natural world. So the Bible, like Copernicus himself, could describe a wondrous "sunset" without error. Rather than reflecting a fundamental conflict between science and Christianity, the Galileo affair is better characterized as the fruit of rapid scientific change, political turmoil, and personal vanity.

Like other warfare myths, distorted accounts of Galileo's encounter with the Catholic Church entered Anglo-American astronomy textbooks in the early nineteenth century, rooted in Protestant polemics against Catholicism rather than in an aversion to Christianity itself.[45] But another myth that creeped into astronomy textbooks in the nineteenth century owed much of its sustaining power to the assumption that Christianity was an irrational hindrance to the progress of science. This was the myth of Copernican demotion, which the next chapter investigates.

6

COPERNICAN DEMOTION

The main objection to Copernicus's views of planetary motion was that they ran counter to the widely accepted teachings of the ancient Greek philosopher Aristotle, whose views were supported by religious doctrine. After all, if the Earth is just one of the planets, it holds no special place in the universe.

    —*Stephen E. Schneider and Thomas T. Arny*, Pathways to Astronomy *(2018), 91*

Humankind has been torn from its throne at the center of the cosmos and relegated to an unremarkable position on the periphery of the Milky Way Galaxy. But in return we have gained a wealth of scientific knowledge. The story of how this came about is the story of the rise of science and the genesis of modern astronomy.

    —*Eric Chaisson and Steve McMillan*, Astronomy: A Beginner's Guide to the Universe *(2017), 26*

M any people believe that Nicolaus Copernicus (1473–1543) demoted humans from the privileged "center of the universe" and thereby challenged religious doctrines about human importance. Atheist Christopher Hitchens even called Christianity's earth-centered cosmology its "greatest failure."[1]

The atheist failed in this case. The beliefs embedded in this narrative about Copernicanism are myths.

Copernicus, a canon in the Catholic Church, considered his heliocentric (sun-centered) astronomy to be compatible with Christianity,
declaring on one occasion that God had "framed" the cosmos "for our
sake."[2] Most other early modern advocates of heliocentric astronomy
similarly affirmed the harmony of the Bible with the new astronomy.

The myth that Copernicus demoted humans assumes that premodern geocentricism (earth-centered astronomy) was equivalent to
anthropocentrism (human-centered ideology). But according to the

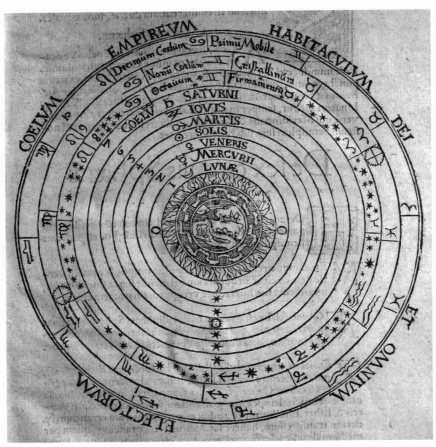

Figure 6.1: Peter Apian's Cosmographia *(1545) depicts the Aristotelian-
Ptolemaic cosmos as almost all European thinkers accepted it up through the
sixteenth century. Earth is at the undignified bottom of the universe. From the
moon (Lunae) upward, the cosmos was thought to be composed of an incorruptible
heavenly element, beyond which is the spiritual heaven. (Photograph courtesy
of the History of Science Collections of the University of Oklahoma)*

ancient Greek geocentric viewpoint that was commonly accepted through the time of Galileo Galilei (1564–1642), earth was at the *bottom* of the universe. This was no honor. "Up" points to the exalted, incorruptible cosmic heaven (see Figures 6.1 and 6.2). "Down" here in the terrestrial realm, things fall apart. C. S. Lewis summarized the medieval vision of the human place in the cosmos to be "anthropope-ripheral" (not human-centered).[3] Accordingly, Galileo wrote in 1610: "I will prove that the Earth does have motion...and that it is not the sump where the universe's filth and ephemera collect."[4] Galileo offered heliocentrism as a promotion for humanity out of the filthy cosmic center, the very center of which, at earth's core, Dante Alighieri (ca. 1265–1321) had identified as the location of hell.

The Copernican reclassification of earth as a planet also invited a new approach to considering the possibility of life on other planets. If

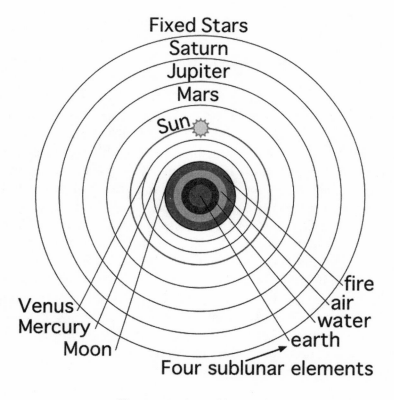

Figure 6.2: Aristotle's cosmology

there is intelligent life on at least one planet (earth), then why not, reasoning weakly by analogy, intelligent life on other planets, too? Kepler thought ET existed but maintained that this did not diminish biblical human significance. In the generation after Kepler, however, it became fashionable in some circles to discuss how Copernican-reinvigorated speculation about ET might revise or undermine Christianity.[5] Thus, in conjunction with such contemplation, the Copernican demotion myth began its long incubation period in popular culture.

The idea that Copernicus demoted humans and thus challenged religion emerged in the mid-seventeenth century as part of an invented anti-Christian narrative. By the mid-nineteenth century the myth had entered astronomy textbooks, and by the second half of the twentieth century it had become textbook orthodoxy.

If anything, the Copernican myth has become more widespread, with scientists and popular writers stretching the myth far beyond what its original proponents had claimed.

The Roots of the Copernican Cliché

Bernard le Bovier de Fontenelle (1657–1757) emerged as the most influential architect of the misconceived Copernican story. Fontenelle's *Discourse on the Plurality of Worlds* (variously titled and translated), which he wrote as an entertaining dialogue about ET within a Copernican framework, remained in print and widely read for a century after its original 1686 French edition. Even after its initial century of popularity, it has occasionally appeared in new printings up to the present. In the first English translation (1688), the character in the story that speaks for Fontenelle declared that it was "well done" of Copernicus to diminish "the Vanity of Mankind, who had taken up the best place in the Universe" (the center), and "it pleaseth me to see the Earth in the crowd of the Planets."[6] Countless kinds of intelligent life, *not* descended from Adam (and thoroughly unlike us), dominated this throng of intelligent creatures in Fontenelle's system of worlds. Humans would be barely noticed in the crowd.

One implication of Fontenelle's playful discussion could be that the uniqueness and cosmic significance of Jesus's earthly incarnation and atonement was a socially constructed exercise in human vanity.

Accordingly, at one point his character blurted, "I know nothing in the World which is not a Monument of the Folly of Man." The historic Christian understanding of the person and work of Jesus Christ (Christology) would be no exception to this rule. But Fontenelle conveniently assumed that his own heterodox viewpoint *was* an exception. Never mind that previous centuries of Christian theological reflection had produced possible ways of reconciling aliens with the Bible—a topic Fontenelle avoided. Fontenelle promoted fashionable doubt about historic Christology while claiming in his book's preface to offer "an Infinity of Worlds with Inhabitants" that is in harmony with "Religion" and even "Scripture."

The literary scholar Dennis Danielson has identified the Copernican demotion myth as the "great Copernican cliché," tracing it back to the mid-seventeenth century, and especially to Fontenelle.[7] Danielson explains:

> Once the center was seen as being occupied by the royal Sun, that location *did* appear to be a very special place. Thus we anachronistically read the physical center's post-Copernican excellence back into the pre-Copernican world picture—and so turn it upside down. But I also suspect (though can't yet prove) that the great Copernican cliché is in some respects more than just an innocent confusion. Rather, it functions as a self-congratulatory story that materialist modernism recites to itself as a means of displacing its own hubris onto what it likes to call the "Dark Ages." When Fontenelle and his successors tell the tale, it is clear that they are making no disinterested point; they make no secret of the fact that they are "extremely pleased" with the demotion they read into the accomplishment of Copernicus.[8]

The Cliché Creeps into Textbooks

About the time American astronomy came of age in the mid-nineteenth century, the Copernican demotion myth began its slow migration into textbooks. In his *Treatise on Astronomy* (1849), Horatio Robinson (1806–1867), a sometime mathematics teacher in the United States Navy, depicted early Copernican advocates as anti-religious warriors:

The true solar system...is called the Copernican system.... But this theory, simple and rational as it now appears, and capable of solving every difficulty, was not immediately adopted; for men had always regarded the Earth as the chief object in God's creation; and consequently man, the lord of creation, a most important being. But when the Earth was hurled from its imaginary, dignified position, to a more humble place, it was feared that the dignity and vain pride of man must fall with it; and it is probable that this was the root of the opposition to the theory.[9]

Fontenelle's tale, after a two-century incubation period, had finally appeared in the American astronomy classroom. Because of its nineteenth-century status as a cultural cliché, Robinson had little reason to check the story's accuracy. Moreover, Robinson (whose personal religious views are unknown) added spin to the story: the alleged demotion became the primary basis for opposition to the theory—a theological objection to human dethronement. Contrary to Robinson, early modern astronomers resisted a moving earth chiefly for scientific, not theological, reasons. During Galileo's career, Tycho Brahe's geoheliocentric system was widely believed to best fit prevailing physical theory and telescopic observations such as the phases of Venus, as explained in chapter 5. Although Isaac Newton's physics displaced Tycho's cosmology, a few Copernican problems remained unsolved until about the time of Robinson's 1849 textbook.[10]

Cardinal Bellarmine's pivotal April 1615 letter, examined in chapter 5, never claimed that Copernicanism challenged human dignity. This leading Catholic theologian wrote in opposition to Copernican astronomy as an unproved theory that seemed difficult to reconcile with biblical descriptions of the sun and earth. Although he had "very great doubts" about the future success of Copernican astronomy, he surely would have ruled it out in principle had he thought that it would demote humanity from its special role as the image bearers of God. But Robinson's 1849 tale of a Copernican challenge to human dignity is nowhere in Bellarmine's letter.

Interestingly, few other astronomy textbooks repeated Robinson's 1849 Copernican demotion story until the second half of the twentieth century, when an extended version of the tale became the dominant textbook narrative. This resurgence of human demotion storytelling occurred during the UFO craze of the 1950s and the subsequent launch of the scientific search for ET intelligence. Out of a sample of fifty astronomy textbooks published between 1849 and 1954, I could find only two that included the false Copernican narrative.

Who were these oddballs? David Todd (1855–1939), known for his attempts to detect communications from Martians, wrote one of these textbooks in 1897 while he was professor of astronomy and director of the observatory at Amherst College. Much later, in 1922, after a decade of increasingly erratic behavior, he was institutionalized for a psychiatric condition caused by unknown factors.[11] The sane Todd of 1897 wrote, "Till the time of Copernicus our abode was generally believed to be enthroned at the center of the universe." But "now we know, what is far less gratifying to our self-importance, that this earth is only one—a very small one, too—of the vast throng of celestial bodies scattered through space."[12] Todd bundled the Copernican demotion and "premodern small universe" myths and embellished them with moralizing rhetoric about humility.

The second textbook in this period to perpetuate the Copernican misconception was written by the director of the Smithsonian Astrophysical Observatory, Charles Abbot (1872–1973).[13] *The Earth and the Stars* (1925), published as part of the Smithsonian Scientific Series, bore the Smithsonian Institution's emblem and motto on its cover. In this textbook, Abbot claimed that Galileo's persecution was linked to the earlier Copernican demotion of earth: "So great a derogation as this of the majesty of the Earth, and consequently of man, its chief citizen, seemed impious in the eyes of most of Copernicus's contemporaries and successors for a century and more; so that we find the immortal Galileo suffering persecution in his old age."[14]

But Galileo's own words were at odds with this Galileo-linked Copernican demotion account. "I will prove," he wrote, that earth moves and that "it is not the sump where the universe's filth and ephemera collect."[15] Although Galileo failed to deliver on the promised proof, he framed his conclusion as a promotion, not a demotion, for earth. Our terrestrial home would no longer be at the *bottom* of the cosmos, which the traditional Aristotelian scientific sense had conceived as something like a sewage container.

Why did Abbot indulge in the rhetorical flourish of calling Galileo "immortal" in the face of church persecution? Why perpetuate the Copernican demotion myth with such enthusiasm in a "Smithsonian Scientific" textbook?

The answer is found, in part, in an essay he wrote offering philosophical lessons in the wake of the infamous 1925 Scopes trial. This was the courtroom extravaganza over biological evolution in American high schools. In his essay, Abbot recommended a liberal view of religion, which meant accepting the "glorious and comfortable" ethical maxims of the Bible while rejecting its "contradictory and barbarous statements" along with its claims of miracles that "the scientific man" deems objectionable.[16] Abbot appears unaware that the idea of miracles actually *depends* on a prior belief in a backdrop of natural regularity against which the miraculous is recognizable, and so miracle and natural law are compatible concepts.[17]

Notice also Abbot's great optimism in the outcome of cosmocultural evolution, thought to be (then) in its early techno-scientific stage. He proclaimed: "May we not anticipate for the future a more wonderful man, whose superhuman attributes and powers we cannot prophesy? Such is the groundwork of the evolutionary faith."[18] According to this naturalistic faith, humans in their lowly origin from terrestrial animals have no exceptional cosmic status, but cosmocultural evolution—building on the "immortal" achievement of Galileo—is likely to evolve us into superintelligent creatures. So the Copernican demotion will lead to our *promotion*, as humanity progresses godlessly toward godlike superintelligence. Although only the demotion segment of this story made its way into Abbot's textbook, the grand promotion narrative has reappeared in recent scientific and technological literature with a vengeance, as we will see in the next chapter.

Four years before Abbot's erroneous textbook story of demotion and persecution, the prominent American astronomer Harlow Shapley (1885–1972) had published a revised version of his high-profile debate with Heber Curtis over the size of our galaxy. The debate took place at the National Academy of Sciences meeting on April 26, 1920. Shapley's story about the perceived meanings of a succession of scientific discoveries went further than the accounts of Todd (1897) and Abbot (1925):

> The physical universe was anthropocentric to primitive man. At
> a subsequent stage of intellectual progress it was centered in a

restricted area on the surface of the earth. Still later, to Ptolemy and his school, the universe was geocentric; but since the time of Copernicus the sun, as the dominating body of the solar system, has been considered to be at or near the center of the stellar realm. With the origin of each of these successive conceptions, the system of stars has ever appeared larger than was thought before. Thus the significance of man and the earth in the sidereal scheme has dwindled with advancing knowledge.[19]

Shapley concluded that our galaxy is "at least ten times greater in diameter" than many others had argued, and that "the solar system can no longer maintain a central position." He became the primary public spokesperson for this *extended* version of the Copernican demotion story that soon made its way into astronomy textbooks, beginning with one written by his most famous female student.

The Extended Copernican Demotion Myth

A century after Horatio Robinson inaugurated the Copernican demotion myth in astronomy textbooks, the Harvard astronomer Cecilia Helena Payne-Gaposchkin (1900–1979) inserted an extended version into her *Introduction to Astronomy* (1954). "The advance of astronomical knowledge," she declared, "has successively dethroned the Earth, the sun, and the stellar system [i.e., our galaxy] from their supposed unique and central stations."[20] She had, no doubt, plenty of opportunity to imbibe this story from her PhD adviser, and subsequent Harvard colleague, Harlow Shapley. As America's celebrity astronomer prior to Carl Sagan, Shapley had launched a pantheistic-based "lifetime assault on anthropocentric thinking"—so a leading Shapley scholar reports.[21] Shapley advanced his progressive Copernican demotion story from the famed 1920 debate through the rest of his career, including his time as mentor, and then colleague, to Payne-Gaposchkin.

The work of another famous cosmologist might have additionally influenced Payne-Gaposchkin to include the extended Copernican demotion in her 1954 textbook. Two years earlier, in his book *Cosmology*, Hermann Bondi (1919–2005) had used the term "Copernican principle" to refer to the idea "that the Earth is not in a central, specially favoured position" in the cosmos.[22] Scientists universally accepted the Copernican principle, he insisted—an attitude produced by the unstated conviction that those who failed to obey "the principle" were simply not scientists. This apparent first-published instance of the term "Copernican principle" further dignified the demotion story and preconditioned scientists to "see" this principle play out in multiple astronomical discoveries, thus contributing to the rise of the *extended* Copernican myth.

Curiously, one of the strongest critiques of the historical claim that the "Copernican principle" contributed to successive scientific discoveries comes from one of Shapley's last students, the astronomer and historian of science Owen Gingerich (now an emeritus Harvard professor). He introduces the topic this way:

> In the twentieth century it has become increasingly popular to refer to a "Copernican principle," namely: We should not consider ourselves to be on a special planet circling round a special star that has a special place in a special galaxy. With respect to the cosmos we should not be considered special creatures, even though we clearly are with respect to life on earth. In full dress, this is the principle of mediocrity, and Copernicus would have been shocked to find his name associated with it.[23]

Gingerich shows that the history of science includes only four episodes in which the Copernican (or mediocrity) principle plausibly *could* have helped advance our scientific knowledge. Of these four cases, he argues, the principle was used only once (and only implicitly in that case). So, he asks, why do so many astronomers turn "feisty" when you challenge the principle? It appears that it is not a scientific principle that aids in astronomical discovery. Rather it is a philosophical preference that is imposed retrospectively on scientific findings that superficially resemble the thematic pattern of the principle. In short, the Copernican-mediocrity account is fatally flawed both historically and scientifically. It is materialist philosophy triumphantly, but erroneously, proclaiming itself to be on the right side of history.

Soon after Payne-Gaposchkin inserted the Copernican demotion into her 1954 textbook, many others followed suit. Some got creative, contributing new components. In his 1964 textbook, Stanley Wyatt

extended the myth *backward* in time by claiming that "when the doctrine of uniqueness was waning," Copernicus "took the ultimate revolutionary step and deposed the earth from its fixed place at the center
of the universe."[24] He also extended the dethronement myth *forward*
in time by including discoveries beyond what Payne-Gaposchkin had
covered. After crediting Shapley for the "dethronement of the sun and
nearby stars from their central position in the galaxy," Wyatt updated
the mediocrity lesson on an intergalactic scale: "Today the seeming
centrality and fixity of our own galaxy is...illusory; the same effect
is seen by an observer in any galaxy. The same is true of motions; the
expansion [of the universe] is recorded by an observer in any galaxy....
The central place of rest in the universe is everywhere and nowhere."[25]

Wyatt's 1964 textbook was written just before Arno Penzias and
Robert Wilson discovered the cosmic microwave background radiation
of the Big Bang. But even prior to this dramatic confirmation of Big
Bang cosmology, observations and arguments for cosmic expansion
made Big Bang cosmology a promising theory. In this intellectual climate Wyatt was determined to use the cosmic expansion of Big Bang
cosmology to promote a Copernican demotion storyline that allegedly
undermined any legitimate belief in a special purpose for humanity.
This move was, and is, controversial, as can be seen by the long and
continued debate over what cosmology implies about humanity.

For example, cosmic expansion within Big Bang cosmology was
pivotal in establishing that the universe had a beginning. And if it had
a beginning and sequence of development, then the present moment
is not like every other moment, and so the Copernican principle fails
at least in regard to time. The first moment of the Big Bang seems
utterly *special* compared to other cosmic slices of time. That clearly
violates the strongest formulation of a Copernican-mediocrity rule.
Moreover, some argue that the cosmic beginning and other features
of contemporary cosmology point to a purposeful cause of the universe.[26] While the purposeful implications of cosmic expansion rest
on a series of reasonable (though highly contested) arguments, the
extended Copernican demotion motif is mostly an emotive story with
little rigorous argumentation.

According to my survey, 71 percent of astronomy textbooks used
in classrooms today perpetuate the Copernican demotion myth—
sometimes laced with explicit (though unwarranted) philosophical

lessons bashing theism. The Copernican misconception has been the dominant textbook narrative since the 1960s, and almost always as the extended version of the myth. Rather than bore you with repetitive quotations from this half-century pile of textbooks, I will leap to textbooks currently in use.

A few current textbooks see a further Copernican demotion for humanity in today's theory of the overwhelming cosmic dominance of dark matter in proportion to ordinary matter, given that human bodies and objects in our daily experience seem to be composed of only regular material elements. Stephen Schneider and Thomas Arny's textbook, for example, asserts:

> It is interesting to consider how far we have moved from our Earth-centered view of the universe in our exploration of galaxies. We have learned that the Sun lies in the outskirts of a galaxy, which is not a particularly significant galaxy, which is in a minor cluster of galaxies. And now we are realizing that the kind of matter that makes up everything we know is just a minor kind of matter in the universe. This is the Copernican revolution taken to extremes![27]

To spin this as a human demotion—even if disconnected from the fake Copernican history—is a subjective game rather than reputable science. It is sure to impress countless students nonetheless. Just as easily and subjectively one could have declared humans unimportant owing to the earlier discovery—before the advent of mysterious dark matter—that our bodies are composed of the ordinary material elements found in the nonliving world. "See, there's nothing special about material me!" Playing the same game, one could invoke either the commonness or rarity of our material composition as grounds for our unimportance. Humans are losers either way: heads or tails.

The Copernican Demotion Translates into Spiritual Promotion

Another recent trend in textbook Copernican storytelling is particularly surprising. Some authors have connected the front-door antireligious demotion to a back-door spiritual promotion.

The Cosmic Perspective (2017), a widely used introduction to astronomy, features a foreword by Hayden Planetarium director Neil deGrasse Tyson, who frames the Copernican achievement as the first of multiple humiliating discoveries. Ironically, humans are still special, Tyson suggests. How so? Because "the cosmic perspective is spiritual—even redemptive—but not religious." Tyson believes that the Copernican demotion story is redemptive because it saves us from religious ignorance: "The cosmic perspective opens our eyes to the universe, not as a benevolent cradle designed to nurture life but as a cold, lonely, hazardous place.... The day our knowledge of the cosmos ceases to expand, we risk regressing to the childish view that the universe figuratively and literally revolves around us."[28]

Tyson offers to replace religion with secular spirituality. He especially recommends the spiritual feelings he has experienced while watching shows at his own Hayden Planetarium: "I feel alive and spirited and connected. I also feel large, knowing that the goings-on within the three-pound human brain are what enabled us to figure out our place in the universe."[29]

Sociologist Elaine Howard Ecklund has found that secular spirituality like Tyson's is common among scientists today. Among spiritually engaged scientists, 22 percent self-identified as atheists, 27 percent as agnostic, and the rest (51 percent) said they were theists. The nontheistic "spiritual" scientists reported experiences of awe in the face of nature.[30] Such ennobling naturalistic spirituality is present in some astronomy textbooks today, and, ironically, it is often connected to the standard dethronement narrative.[31]

Dennis Danielson analyzes this spirituality as it is commonly justified by a series of human dethronements, beginning with Copernicus:

> But the trick of this supposed dethronement is that, while purportedly rendering "Man" less cosmically and metaphysically important, it actually enthrones us modern "scientific" humans in all our enlightened superiority. It declares, in effect, *"We're truly very special because we've shown that we're not so special."* By equating anthropocentrism with the now unarguably disreputable belief in geocentrism, such modern ideology manages to treat as nugatory or naive the legitimate and burning question of whether Earth or Earth's inhabitants may indeed be cosmically

special. Instead it offers—if anything at all—a specialness that is cast in exclusively existential or Promethean terms, with humankind lifting itself up by its own bootstraps and heroically, though in the end pointlessly, defying the universal silence.[32]

This narrative of dethronement as *subtle enthronement* is pervasive in popular astronomy literature. Consider Eric Chaisson and Steve McMillan's introductory astronomy textbook:

> Yet there was a time, not so long ago, when our ancestors maintained that Earth had a special role in the cosmos and lay at the center of all things. Our view of the universe—and of ourselves—has undergone a radical transformation since those early days. Humankind has been torn from its throne at the center of the cosmos and relegated to an unremarkable position on the periphery of the Milky Way Galaxy. But in return we have gained a wealth of scientific knowledge. The story of how this came about is the story of the rise of science.[33]

The expression "unremarkable position on the periphery of the Milky Way Galaxy" is worth pondering. It shows how the extended Copernican demotion myth often influences scientific assessment of earth's significance today. Since the astrobiologists Guillermo Gonzalez, Donald Brownlee, and Peter Ward introduced the concept of a "galactic habitable zone" (GHZ) in a 2001 paper, studies have largely confirmed this restricted life-friendly zone within a galaxy.[34] Most astronomy textbooks now discuss the GHZ and acknowledge that a planet near a galaxy's center (or too distant from the galactic center) would, owing to multiple factors, be uninhabitable. Curiously, most textbooks still depict our knowledge of earth's noncentral location within our galaxy as a humiliating demotion. The demotion narrative diminishes the significance of our location in the GHZ with no appeal to evidence. Ironically, the human promotion was supposed to involve a "wealth" of new "scientific knowledge." Astronomical "knowledge" without appeal to evidence is not a reasonable upgrade.

So the Copernican demotion myth actually distorts our assessment of current astronomical evidence. This reinforces the importance of identifying and dismantling myths scientists believe.

Chaisson and McMillan promote a naturalistic worldview on the authority of science. Rarely do current textbooks convey naturalism with obvious spiritual overtones the way Neil deGrasse Tyson did in his textbook foreword. Naturalistic spiritualism is still largely in its incubation phase before its possible full outbreak in astronomy textbooks.

To get a better feel for the increasingly popular search for "unlimited value" spirituality within a naturalistic worldview, we could do no better than examine a recent NASA book aimed at public outreach. After essays by leading astrobiology scholars such as Eric Chaisson and Steven Dick, NASA scientist Mark Lupisella (who coedited this book with Dick) writes:

> *A Cosmic Promotion?* Scientists and thinkers have been fond of pointing out humanity's "great demotions." From Copernicus to modern day cosmology (perhaps with the exception of "anthropic principles" and associated observations of "fine tuning"), humanity has been displaced and demoted from privileged positions in the cosmos. Perhaps it's time for a promotion—one that goes beyond the confusion of anthropic principles, one that does not rely on teleological assumptions and assertions about the ultimate nature of the universe. Bootstrapped cosmocultural evolution allows for the possibility that life, intelligence, and culture could have arisen by chance, while at the same time asserting that such phenomena are cosmically significant. Stronger versions suggest that cultural evolution may have unlimited significance for the cosmos. Our cosmic location and means of origin should not be confused with our cosmic potential.[35]

Will this enthusiastic vision of "our cosmic potential" overwhelm the dethronement narrative that has long dominated astronomy textbooks? Jeffrey Bennett's *The Cosmic Perspective* (2017) already has embraced Tyson's Sagan-like mantra: "The cosmic perspective is spiritual—even redemptive—but not religious." Compared to the medieval, Copernican, and Keplerian accounts of cosmic human dignity rooted in Christianity, such twenty-first-century proclamations of the "redemptive" effect or "unlimited significance" of cosmocultural evolution have abandoned cosmic modesty.

This mythic human exultation should not be shocking in the light of mythologist Gregory Schrempp's study demonstrating that popular science writing (including introductory textbooks and the NASA book above) often tries to "show that reality as envisioned by science compensates for the loss of the fantasies allegedly offered by myth—although this is often accomplished, I argue, through the construction of a new myth." Schrempp concludes, "In various forms and degrees of explicitness, some sort of 'trade-in' offer is a recurrent theme in this genre."[36]

Copernican Demotion Traded for Alien Promotion

The history of mythic "trade-ins" extends far back. Early in human history we invented anthropomorphic deities (gods constructed in our own image) to bootstrap our significance and then projected this into our cosmology, making it geocentric and anthropocentric (ignore for a moment that the Judeo-Christian tradition does not fit this storyline). We then, so the story goes, traded this religious superstition for the Copernican worldview—a demotion that gave us modern science as a reward for our newborn humility (likewise ignore how historians of science have debunked this tale). As we tasted the delightful fruit of science, it opened our eyes to the coming enlightening encounter with ET (and/or artificial intelligence), which might usher in the climactic trade-in offer. Cosmocultural evolution promises to generate "cosmic value" of "unlimited significance" (thanks to NASA for this revelation), which we would enjoy after joining the great cloud of extraterrestrial participants of the emerged superintelligence—whether biological or postbiological. Notice the U-shaped story structure: initial spiritual greatness, humiliation, and finally unlimited heavenly exultation. Sound familiar? A new secular Bible story is emerging.

How are these mythic trade-in offers packaged in astronomy education today? First observe the connection between the Copernican demotion and the search for habitable extrasolar planets. Jeffrey Bennett and colleagues write in their textbook:

> The very idea of planets around other stars, or extrasolar planets for short, would have shattered the world views of many peo-

ple throughout history. After all, cultures of the western world long regarded Earth as the center of the universe, and nearly all ancient cultures imagined the heavens to be a realm distinct from Earth. The Copernican revolution, which taught us that Earth is a planet orbiting the Sun, opened up the possibility that planets might also orbit other stars.[37]

In the next chapter I will show that the supposed religion-crushing implications of contact with ET are largely hype that prepares the way for anticipated ET enlightenment—a whopping mythic trade-in offer. For now, just note the Copernican-ET association.

In the textbook *Foundations of Astronomy*, Michael Seeds and Dana Backman tighten up this association and suggest how it might shape expectations of future discovery:

> Why do astronomers seem so confident that there is life on other worlds? No message has been received from distant worlds and no life has been detected on any of the worlds visited by probes from Earth. Yet astronomers face this absence of evidence with confidence because of the work of [Copernicus]. . . . Astronomers have adopted the Copernican Principle: Earth is not in a special place. That principle can be extended to assuming that Earth also is not special in other ways. If Earth is not special, if it is just a planet, then there should be lots of planets like Earth. . . . For astronomers accepting the Copernican Principle, it seems inevitable that life will arise on planets where conditions permit it, and then evolve to become more complex.[38]

So ET can be deemed *inevitable* because of a "Copernican Principle" constructed by humans (ironically, not Copernicus). At least Seeds and Backman implicitly acknowledge that some astronomers, on scientific grounds, dissent from this party line. Such diversity of expert opinion is unsurprising, given the largely philosophical (and pseudo-historical) status of the Copernican principle.

The next chapter examines the most extraordinary "trade-in" offer to arise in popular science writing since the Copernican demotion took hold: ET enlightenment. Are brilliant aliens headed our way now?

7

EXTRATERRESTRIAL ENLIGHTENMENT

There is almost certainly no civilization in the galaxy dumber than us that we can talk to. We are the dumbest communicative civilization in the galaxy.
   —*Carl Sagan, in* Life Beyond Earth and the Mind of Man *(1973), 63*

Any sufficiently advanced technology is indistinguishable from magic.
   —*Arthur C. Clarke,* Profiles of the Future *(1973), 21*

Your merchants were the great ones of the earth, and all nations were deceived by your sorcery.
   —*Revelation 18:23, ESV*

Two leading space science advocates, America's Carl Sagan and Britain's Arthur C. Clarke, agreed in 1973 that humans are dumb compared to any aliens that we would potentially meet. Alien technology, including its artificially intelligent robots, would appear magical to us.

More than two decades later, cosmologist Paul Davies added, "It's inevitable that if we discover life elsewhere in the Universe, it will change forever our perspective of our own species." He specified that "those people who cling to the idea that humanity is the pinnacle of creation, or that somehow we were made in the image of God, would I think receive a rude shock."[1]

Davies and many other scientists anticipate that an encounter with ET will end human-centeredness and cosmic loneliness, ushering in an age of universal spirituality beyond sectarian terrestrial religion. Of all religions, Christianity would be most decisively defeated by this event, because of its cosmically untenable doctrine of the unique incarnation and redemptive work of God's Son on earth—or so the ET enlightenment story goes. Would you trade in your religion (or none) for this vision of the future?

The historic Copernican achievement, and especially its later mythical reinterpretation as a demotion for humanity, helped prepare the way for the futuristic ET enlightenment myth. The Copernican reclassification of earth as a planet (revolving around the sun) brought a new wave of speculation about life on other celestial objects. Kepler thought ET existed, but he maintained, along with many other scientists since, that encountering ET would not undermine Christianity (see chapter 8).

As the Copernican demotion myth became the dominant narrative of twentieth-century astronomical literature, and as its use in anti-Christian rhetoric increased, a craving for a new understanding of human significance also intensified. The ET enlightenment myth now fills that void for many people—as does a companion myth about exponentially increasing artificial superintelligence.

SETI Science and SETI Religion

In 1960, during the postwar UFO fad, astrophysicist Frank Drake helped launch a more serious, scientific search for ET intelligence (SETI) movement. SETI attempts to detect ET radio signals from outer space. In a 2010 jubilee interview with the SETI Institute, Drake said he experienced "a very powerful emotion" in 1960 when apparently he detected communication from ET—although he realized soon thereafter that he had *not* received ET's call. Many scientists get quite emotional about detecting ET because, Drake reported, "you sense that what you're seeing is going to change all of history, and I think for the better."[2] How optimistic is Drake? He thinks ET technology will probably include a recipe for immortality—without God.[3]

After flirting with this idea that aliens will bring technological magic, and even immortality, Paul Davies has recoiled.[4] He concludes that such speculation

> needs to be tempered with a healthy skepticism. There is no doubt that twenty-first-century science is incomplete and provisional, yet it still represents the most reliable approach to knowledge, with the wealth of understanding and experience accumulated over several centuries of careful investigation. In the search for alien intelligence, it is well to adopt a pragmatic view and go with our current picture of science as the best there is so far on offer to guide us, while being open-minded about the possibility of surprises ahead.[5]

Davies argues that scientific restraint defeats ideas such as aliens' traveling (or communicating) faster than the speed of light or over-coming the second law of thermodynamics.[6] Remember, technology can *never* overcome the laws of nature.

Nevertheless, Skeptics Society founder Michael Shermer contin-ues to flaunt his 2002 *Scientific American* "Skeptic" column announc-ing Shermer's Last Law: "Any sufficiently advanced extraterrestrial intelligence is indistinguishable from God."[7] Shermer, the profes-sional skeptic, could use a dose of Davies's "healthy skepticism." So could New Atheist Richard Dawkins, who believes that "there are very probably alien civilizations that are superhuman, to the point of being god-like in ways that exceed anything a theologian could pos-sibly imagine."[8]

Are Shermer and Dawkins ignorant of centuries of philosophi-cal and theological reflection about God as the greatest conceivable being? Such a being, if he exists, would surely outrank *any* ET or *any* technology. To be reasonable and honest, they would need to acknowledge at least this much about the very idea of God in the Judeo-Christian tradition.

Despite the evidence that makes contact with ET very unlikely, science distorted by naturalistic philosophy can make such contact seem very likely, even inevitable. Davies attributes the uptick in alien contact hopes among scientists since 1960 to "fashion rather than discovery." He warns that "we must never allow speculation to replace real science" in SETI.[9]

SETI Science and Science Fiction

Given the vast distances that separate solar systems, any creature able to travel to earth from another habitable planet would have to be vastly superior in technology—so superior that currently only science fiction renders such a trip a compelling story.[10] Noting the silence SETI has so far observed, physicist Stephen Webb points to a comment the science fiction author David Brin made in 1983: "Few important subjects are so data-poor, so subject to unwarranted and biased extrapolations—and so caught up in mankind's ultimate destiny—as is this one." More than three decades after Brin published his analysis of the Great Silence, little has changed. The subject is *still* data-poor.[11]

Even communicating with ET remotely is quite unlikely. For one thing, the communication would probably amount to a mere monologue, rather than an enlightening dialogue, for any one generation of humans, given the vast interstellar distances and thus the lengthy time needed to transmit messages. Harvard astrophysicist Howard Smith notes that the speed of light, which marks the speed limit for both travel and communication in the cosmos, severely limits any plausible alien contact or meaningful communication to a tiny patch of the universe close to earth. Smith summarizes the "cumulative results from exoplanet studies" (and other analyses) this way: "*we are most probably alone*—at least there is probably no other intelligent life within 100 generations' reach to talk with." He adds: "For all intents and purposes, we and our descendants for at least 100 generations are very likely living in solitude. I call this the Misanthropic Principle."[12]

Consider the requirements for a single round of communication (one message received and answered) to plausibly occur within one hundred human generations. With the speed of light representing the maximum speed for communication, we would have to restrict our potential ET communication partners to within a radius of only 1,250 light-years from earth.[13] Within that radius there are about twenty million stars.[14] That may seem like a huge number of solar systems with possible habitable planets, but the large number of "just-right" parameters necessary for life-suitable habitats quickly undercuts (by *multiplying each* improbable factor) the comparatively meager probabilistic resources of those twenty million spins of the local cosmic lottery.[15]

"Just-right" factors include the right location within a galaxy, right kind of host sun, right distance from host sun, right orbital relations to large Jupiter-like planets, right kind of protection from being hit in a life-destructive way by space objects (e.g., asteroids and comets), right kind of protection from harmful radiation, right kind of reception of life-friendly radiation, and right amount of liquid water. In many cases the right *timing* of these and related factors (e.g., migration of Jupiter-like planets within a planetary system) are crucial for life, especially complex life. Many of the items on this list are interlinked in complicated ways, and so there is no simple formula for calculating the overall probability of getting a habitable planet. This much is clear: the probability of getting everything lined up in a life-friendly manner is exceedingly small.[16]

Should we increase the cosmic search radius beyond 1,250 light-years to improve the odds of ET's existence? Well, again, the speed limit of light drastically undermines the likelihood of communication or contact. In fact, as the cosmic search radius increases, we reach *complete* communication isolation from ET, because most galaxies are receding from us at *accelerating* rates. Smith explains how "light signals sent from Earth today will *never* even catch up to galaxies whose light has taken only about 10 billion years to reach us." These galaxies, though "well within our cosmic [visibility] horizon," are "now beyond our [communication] reach forever."[17] With each passing moment we become communicatively isolated from *more* of the cosmos. The hope of ET enlightenment grows dimmer by the second. Time exacerbates loneliness.

Martin Rees, the UK's Astronomer Royal, had some interesting things to say in 2015, soon after becoming chair of the $100 million SETI Breakthrough Listen project funded by Russian billionaire Yuri Milner. Rees expressed confidence that ET "is likely to have long ago transitioned beyond the organic stage" to the "more powerful intellects" of machine (artificial) intelligence. So SETI "ought to be looking" to nonearthlike environments: "Interplanetary and interstellar space is where robotic fabricators will have the grandest scope for construction, and where non-biological 'brains' may have insights as incomprehensible to us as string theory is to a mouse."[18] Since the vast majority of the universe is environmentally nonearthlike, Rees's vision would seem to set an enormous search task for the Breakthrough Listen project. Perhaps that is why the project website prioritizes earthlike environments, despite Rees's claim that humans "ought to be looking" elsewhere.

Around the same time he made his inaugural statements about the Breakthrough Listen project, Rees undercut his hopeful comments about ET by endorsing a more pessimistic take. Rees wrote the foreword to (and highly recommended) the second edition of Stephen Webb's book *If the Universe Is Teeming with Aliens... Where Is Everybody?*, whose title conveys the paradox physicist Enrico Fermi famously expressed. In the foreword, Rees noted: "Maybe we will one day find ET. On the other hand, this book offers 75 reasons why SETI searches may fail." In the seventy-fifth reason, Webb ventures his own solution to Fermi's paradox: that extraterrestrial "sentient, intelligent, sapient creatures that build civilizations and with whom we can communicate" simply "don't exist." Webb's critiques of the other seventy-four explanations point (indirectly) to his own pessimistic solution. Webb's book is one long argument against the "gut reaction we perhaps all feel when we look at the night sky—that there must be intelligent life somewhere out there." Instead, Webb maintains, we "have to be guided by reason, not gut reaction."[19]

Like Rees, Belgian philosopher Clément Vidal is an advocate of ET artificial superintelligence. He claims that humans "are in an extremely short transition phase from technical impotence to technical omnipotence"—a transition almost certainly made multiple times by other races of intelligent beings out there (so he thinks).

Technological omnipotence? This is a modern myth, a new Almighty God.

Even if we assume that such artificial superintelligence is possible (we will assess arguments otherwise), why does Vidal think that finding brilliant ET is so important for humanity? "The main 'objective' or scientific impact of finding ETs will be to literally universalize our knowledge," he writes. "For now, only physics and some chemistry have been proved to hold beyond the confines of our atmosphere." Yet he adds that, "thinking in an astrobiological or cosmological context, authors have already started to universalize many domains of knowledge such as language, sociology, economy, ethics, laws, aesthetics, theology, culture, or eschatology."[20] So the ET enlightenment story already does the heavy lifting in academia before any confirmed ET contact. That is what myths do—imaginative archetypal stories shape how humans think. The ET enlightenment myth has arisen alongside its ancient siblings: the myths of the world's great religions.

Cheerleaders for alien robotic superintelligence, like Vidal, are yearning for a nontheistic theology with its own secular eschatology: prophecies of future technological paradise. But given the highly questionable assumptions that they make about exponential growth in artificial intelligence, much of their quest is driven by pseudo-science.

Likewise, philosopher Susan Schneider confesses her belief that "alien civilizations will tend to be forms of [artificial] superintelligence: intelligence that is able to exceed the best human-level intelligence in every field—social skills, general wisdom, scientific creativity, and so on."[21] She, and numerous others in this SETI camp, avidly promote Arthur C. Clarke's opinion that such advanced alien civilizations have technology that, to us, will be "indistinguishable from magic." Although the arguments for such conclusions about artificial superintelligence are weak, growing belief in this secular ideology leads people to expect the experiential equivalent of occult phenomena.[22]

C. S. Lewis anticipated this bizarre trend in his space trilogy, especially in the last novel, *That Hideous Strength* (1944). Ironically, several critics have complained about the scientifically unrealistic occult content of Lewis's space trilogy—it is not scientific enough to be robust sci-fi, they suggested.[23] And yet Lewis pointed a prophetic finger at the occult-like magical SETI expectations of our own time.

Steven Dick, a leading historian of ET speculation, goes further in identifying the religious impulse that motivates much of SETI:

> It may be that in learning of alien religions, of alien ways of relating to superior beings, the scope of terrestrial religion will be greatly expanded in ways that we cannot foresee. It may be that, as a search for superior beings, the quest for extraterrestrial intelligence is itself a kind of religion.... It may be that religion in a universal sense is defined as the never-ending search of each

civilization for others more superior than itself. If this is true, then SETI may be science in search of religion, and astrotheology may be the ultimate reconciliation of science with religion.[24]

The Technological Singularity: Homegrown AI Magic

The science and religion of SETI, which includes speculation about vastly superior alien artificial intelligence, is related to the increasingly popular expectation that artificial intelligence (AI) produced by *terrestrial* scientists and engineers will soon overtake humanity. According to this view, AI will become exponentially superintelligent and reach a tipping point called the "technological singularity." Many Singularitarians, as they are called, even claim that humans will achieve immortality and superintelligence when we merge with this future machine intelligence. This is the terrestrial AI companion to the ET enlightenment myth.[25]

The most sophisticated and charming defender of "the Singularity" is Ray Kurzweil, who is sometimes called the Thomas Edison of AI. Kurzweil has invented many useful gadgets run by sophisticated software, and Google appointed him chief engineer in 2012.

In 2005 Kurzweil said that the magical Harry Potter stories "are not unreasonable visions of our world as it will exist only a few decades from now," when the Singularity arrives.[26] Thanks to artificially intelligent technology, he claimed, "the entire universe will become saturated with our intelligence."[27] When that happens, the cosmos will "wake up" and "be conscious, and sublimely intelligent."[28] Everything is evolving toward "infinite beauty, infinite creativity, infinite love."

So the universe inevitably will become an entity that Kurzweil, a secular Jew, thinks resembles the traditional monotheistic idea of God. We have, yet again, reached the Olympian heights of myth, but one that makes Zeus look like a kindergartner playing in a small sandbox. Jump on board this deification process by uploading your mind to supercomputers that will be available (hopefully) before most of us die. Meanwhile, begin taking Ray Kurzweil's nutritional supplements and other "steps to living well forever" at rayandterry.com. Could you lose your (potential) salvation if you fail to take some of the countless supplements that Kurzweil himself swallows daily?[29]

Murray Shanahan, a leading British advocate of the Singularity, downplays Kurzweil's techno-religious sensationalism. Nevertheless, as an AI expert, he predicts (echoing Arthur C. Clarke) that "human-level AI, however it is achieved, is likely to lead directly to technology that, to the rest of us, will be indistinguishable from magic." All it requires is "faster computation."[30] He sees this technological "magic" coming in a near-future Singularity.

Singularity Skepticism

The ET enlightenment story, with its AI component, is a myth in both senses of the word used in this book—an imaginative archetypal story, and a false story. To understand why, we must understand what AI is, and what the barriers to runaway artificial superintelligence are.

One of the leading Singularity skeptics is the Australian AI scientist Toby Walsh. He thinks that meta-intelligence considerations—that is, the higher-order ability to *improve* your intelligence—constitute "one of the strongest arguments" against the Singularity. According to this argument, Singularity believers confuse "intelligence to do a task with the capability to improve your intelligence to do a task." Walsh explains:

> Suppose an AI system uses machine learning to improve its performance at some tasks requiring intelligence like understanding a text, or proving mathematical identities. There is no reason that the system can in addition improve the fundamental machine learning algorithm used to do this. Machine learning algorithms frequently top out at a particular task, and no amount of tweaking, be it feature engineering or parameter tuning, appears able to improve their performance.[31]

While "not predicting that AI will fail to achieve super-human intelligence," Walsh suggests "that there will not be the run-away exponential growth predicted by some." AI-singularity salvation stories require such exponential growth, and more.

Meet Margaret Boden, another "S-skeptic," as she calls herself. She is a cognitive scientist at the University of Sussex. There she helped develop the world's first academic program devoted to understanding how the human mind works by means of philosophy, artificial intelligence, neuroscience, linguistics, and other social sciences. To assist in the interdisciplinary endeavor, she has degrees in the medical sciences, philosophy, and psychology. In her research, she integrates these disciplines with AI. Boden grounds her skepticism in the fact that artificial *general* intelligence (AGI) "is proving so elusive."[32] AGI stands between current AI and the presumed artificial superintelligence that some believe will arise in the Singularity. Good examples of current AI can be found in Apple's Siri and IBM's Watson, which serve and amuse humanity in narrowly structured ways. What would it take to achieve even just AGI, which is AI that "learns" much more universally or generally than Siri or Watson? It would require a more thorough solution to what AI thinkers call the "frame problem." The frame problem is, roughly, the challenge of finding adequate collections of axioms for a viable description of an AI device's environment, to give that device the appropriate frame of reference for solving problems.[33]

Boden explains that human thinkers tacitly assume certain implications, whereas a computer ignores the same implications because they "haven't been made explicit." AI systems "lack common sense," Boden writes. AI scientists such as Shanahan "sometimes claim that the frame problem has been solved" because the problem "can be avoided if all possible consequences of every possible action are known."[34] But that solution does not address the *general* frame problem, which still "lurks all around us—and is a major obstacle in the quest for AGI."[35] And if you want artificial superintelligence, you need AGI first.

Singularitarians "ignore the limitations of current AI," Boden says, because they believe "they have a trump card: the notion that exponential technological advance is rewriting all the rule books."[36] Boden rejects this unreasonable extrapolation—it leaves people wide open to the mental habits typical of the occult, as C. S. Lewis warned.

Even increases in computational power and data availability "won't guarantee human-like AI," Boden writes. Thomas Dietterich, former president of the Association for the Advancement of Artificial Intelligence, puts Boden's point this way: Most AI research groups are "focused on improving computer performance on narrow tasks. It is easier to make progress and measure success on narrow tasks, whereas it is difficult to develop useful measures of general intelligence. We hardly know where to begin."[37]

AI scientist and entrepreneur Erik Larson has probed deeper into the disconnect between machine computational "learning" and true human learning. Much of human learning goes far beyond inductive generalizations made by gathering and computationally processing data. "In fact," Larson writes, "there is no computational framework for epistemologically complex learning at all—if we could do that, we could solve the frame problem." Humans reason in different ways, and Larson notes that AI "has made progress only towards problems that admit of computational representation and processing, ones that importantly presuppose a particular simplistic set of epistemological conditions."[38]

What about when an object or event can be explained in multiple possible ways? Such situations frequently come up in science and even in everyday human reasoning. Why does my car not start? Any number of explanations could apply, and to find the best explanation— the theory that is *probably* true—you need to assess a range of traits known as theoretical virtues: evidential accuracy, causal adequacy, explanatory depth, internal consistency, internal coherence, universal coherence, beauty, simplicity, unification, durability, fruitfulness, and applicability.[39] With good reason these traits are called "virtues," not "rules." Most of them lie far beyond the rule-governed reach of AI. As Larson notes, "there's no blueprint" for converting AI's dataset analysis to an assessment of so many different virtues when multiple explanations seem plausible. He adds: "No one who is honest has the faintest clue what such a conversion would even look like. It's simply a mystery."[40]

Singularity skeptics Alessio Plebe and Pietro Perconti make a related point about aspects of human intelligence that appear to be far beyond AI emulation. In recent decades cognitive scientists have discovered that human intelligence is much more multifaceted than previously thought. They have identified many kinds of intelligence and are proposing still more, "from emotional to musical, from spatial to social." So "the number of aspects one has to take into account" multiplies with each new advance in cognitive science. Consequently, research in AI "heads more towards a slowdown rather than towards a singularity effect."[41]

Although alien intelligence, if it exists, might differ significantly from how our minds work, any technologically capable intelligence in our universe is sure to be multifaceted in ways that AI cannot achieve.

Similar to the "slowdown" worry of Plebe and Perconti, Toby Walsh identifies the "diminishing returns argument" against the Singularity. Walsh notes that the Singularity supposes that AI improvements "will be a relative constant multiplier, each generation getting some fraction better than the last."[42] In practice, however, "the performance so far of most of our AI systems has been that of diminishing returns." Walsh explains, "There is often lots of low hanging fruit at the start, but we then run into great difficulties to improve after this."[43] When Kurzweil and other early AI researchers quickly solved many of the easier problems, they made overly optimistic projections of future AI progress. Moreover, as we deepen our understanding of complex topics like AI, the subject matter becomes more difficult, requiring more and more specialized knowledge."[44] Microsoft cofounder Paul Allen calls this the "complexity brake." Walsh concludes: "Even if we see continual, perhaps even exponential improvements in our AI systems, this may not be enough to improve performance. The difficulty of the problems required to be solved to see intelligence increase may themselves increase even more rapidly."[45]

Andrew Majot and Roman Yampolskiy likewise argue that AI improvements are bound by the inherent limits of algorithm complexity (an algorithm is a set of rules that enables calculations or other problem-solving procedures in computers). They note that "the law of diminishing marginal returns would probably take effect, causing greater and greater computational resource usage for smaller and smaller performance gains."[46]

Plebe and Perconti comment further in this vein: "As far as we know, there is no example of any artificial system able to design new algorithms."[47] What computer scientists call "automatic code generation" does not constitute computers' designing new algorithms by themselves. Rather, human programmers use such tools to write code. Plebe and Perconti conclude from all this that "so far, no sign of the aspect of intelligence that is most crucial to the singularity hypothesis has been seen yet" in AI.[48]

For all these reasons (and others), exponentially increasing artificial superintelligence appears prohibitively unlikely. Note, too, how these barriers surveyed below are not peculiar to humanity or our cosmic location. As such, these arguments against the Singularity radically undercut both the expectation of ET enlightenment and the terrestrial dreams of futurists such as Kurzweil.

Incidentally, Kurzweil was not the first to suggest salvation by mind upload. Hans Moravec of the renowned Carnegie Mellon University Robotics Institute "effectively began the Apocalyptic movement in 1978" with an essay that appeared in the sci-fi magazine *Analog*.[49] He imagined a way to upload human minds to cyber-immortality with the help of future AI robotic surgeons, thereby updating a similar story that had appeared in Arthur C. Clarke's sci-fi novel *The City and the Stars* (1956). Unlike Clarke, Moravec is an AI pioneer and thus often perceived as more credible. But few AI experts actually swallow the mind-uploading apocalyptic AI story. Margaret Boden decries the "near-absurdity" of such "transhumanist philosophy."[50] MIT's famous linguist Noam Chomsky labels the Singularity "science fiction."[51] Steven Rose, neurobiologist at England's Open University, calls the uploaded consciousness scenario "pretty much crap."[52]

Enlightenment from Heaven or Earth?

Kurzweil promises human-engineered everlasting bliss in the post-Singularity world when our minds have been uploaded to computers. Paul Davies, as noted, offers an alternative myth. He has set his hope on a superior spirituality that will come down from the heavens when biological or robotic ET enlightens us. Which will it be, the computational rapture of the terrestrial nerds, or ET salvation? Or neither? People long to be "caught up in the cloud" one way or another. Kurzweil and Davies are key proponents for these two visions of our future. Kurzweil thinks humans are the only technological creatures in the universe.[53] Davies imagines humans as a marginal species in a crowd of morally superior extraterrestrial overlords.

What makes Kurzweil think we are alone, thus undercutting heavenly enlightenment hopes? He writes: "Once a planet yields a technology-creating species and that species creates computation (as has happened here), it is only a matter of a few centuries before its intelligence saturates the matter and energy in its vicinity, and it begins to expand outward at at least the speed of light."[54] Kurz-weil argues that by now we would have detected at least one of the countless ET civilizations predicted by SETI enthusiasts (if they existed), given that most of them would have had millions of years to expand their omnidirectional technological signature of existence at the speed of light. But what if these alien civilizations are deliberately keeping us in the dark about their existence? Or what if they destroy themselves by nuclear holocaust before we detect their technological signatures? Kurzweil responds to arguments like these by noting that it is not credible to assume that *all* of the billions of ET civilizations postulated by SETI scientists would have succumbed to one or more of these modes of cosmological silence.[55] So ET is extremely unlikely to exist, he concludes.

In this context Kurzweil compares three views of the human place in the cosmos: pre-Copernican, Copernican-SETI, and Copernican-Singularitarian:

> Our naive view of the cosmos, dating back to pre-Copernican days, was that the earth was at the center of the universe and human intelligence its greatest gift (next to God). The more informed recent view is that, even if the likelihood of a star's hav-ing a planet with a technology-creating species is very low (for example, one in a million), there are so many stars...that there are bound to be many...with advanced technology. This is the view behind SETI—the search for extraterrestrial intelligence—and is the common informed view today. However, there are [Singularitarian] reasons to doubt the "SETI assumption" that ETI [extraterrestrial intelligence] is prevalent.[56]

Curiously, Kurzweil's own Singularity doctrine, unlike the once-dominant pre-Copernican European cosmology, is *actually* rather than *only apparently* human-centered in an ideological sense:

> But it turns out that we are central, after all. Our ability to cre-ate models—virtual reality—in our brains, combined with our modest looking thumbs, has been sufficient to usher in another form of evolution: technology. That development enabled the persistence of the accelerating pace that started with biologi-

cal evolution. It will continue until the entire universe is at our fingertips.[57]

Davies sees the situation differently. He says that human technology will seem primitive compared with that of ET. Nor will technology mark the only difference. Davies argues it is "likely that any civilization that had surpassed us scientifically would have improved on our level of moral development, too." That is why he believes that when biological or robotic ET arrives, humans will see a need to upgrade or discard traditional religions. He particularly targets Christianity. "Can we contemplate a universe that contains perhaps a trillion worlds of saintly beings, but in which the only beings eligible for salvation inhabit a planet where murder, rape, and other evils remain rife?"[58]

Here Davies crucially misrepresents what many thoughtful Christians have said about this topic. For example, C. S. Lewis observed that it would be a mistake to think the work of Christ on earth "implies some particular merit or excellence in humanity. But of course it implies just the reverse: a particular demerit and depravity. No creature that deserved redemption would need to be redeemed."[59] Lewis even imagined other worlds with sentient creatures who never rebel against God and thus do not need the sacrificial work of Jesus on their behalf. In short, ET and Christianity might be compatible.

But not all ET-AI scenarios are Christian-friendly. Kurzweil's and Davies's views have the potential to create a global alliance with many of the world's religions—with the notable exception of Christianity. Imagine that a *seemingly* conscious and superintelligent machine appears on earth. Now imagine that it solves some of the world's toughest problems. Suppose, as predicted by the ET-AI enlightenment myth, some of this creature's capabilities appear magical, such as producing "immortality" by consciousness upload (for the terminally ill and others declared "ready" to upgrade beyond embodied humanity). Many religious leaders might champion this apparently benevolent and miraculous creature. Singularity advocates probably would opt for a naturalistic terrestrial origin theory of the creature and announce that the Singularity has arrived. SETI enthusiasts would favor a naturalistic extraterrestrial origin theory: alien machine superintelligence. Paul Davies places his bet on this and foresees worldwide religious reorientation in response to the alien oracle. Despite the radically different viewpoints within this scenario, most people probably would follow the creature if it promoted prosperity and techno-spirituality. This widespread conversion, however, would be impossible for Christians unless they abandoned historic Christian beliefs about God, the cosmos, humanity, and redemption.

ET and AI Technological Salvation: United Against Christianity

The futuristic ET enlightenment myth and its terrestrial Singularity companion share a strong aversion to Christianity. This is similar to the six historical myths covered in the previous chapters. Those historical myths erroneously assert a state of necessary warfare between science and Christianity.

Both ET and AI salvation are supposed to occur by technological good works, not by accepting the undeserved sacrificial gift of Jesus. Although ostensibly secular, the ET and Singularity ideologies share what Steven Dick has called "religion in a universal sense," which he takes to be "the never-ending search of each civilization for others more superior than itself." ET and AI salvation differ mainly in whether separately evolved aliens travel here to enlighten us or *we become the aliens* through an electronic transhumanist reconstruction (consciousness upload) that eventually fills the universe with intelligence and enlightenment. Placing their hope in salvific technological enlightenment, both ideologies reject the idea of the Judeo-Christian God from whom humanity has been tragically alienated.

The ET-AI salvation story is seriously lacking in support. ET might exist and AI might become extraordinarily advanced, but neither is plausibly our savior. They are futuristic myths bearing unreasonable technological promises of worldview-changing enlightenment and immortality.

Here we see how myth as "imaginative archetypal story," rather than merely "false story," can be past-oriented *or* future-oriented. Classical mythology is *past*-oriented, focusing on the origins of peoples and their gods. Part two of my book begins by exploring how sci-fi often functions as *futuristic* myth, and how this influences scientists and pop culture.

Part 2

SEEKING SOURCES AND SOLUTIONS FOR THE WARFARE MYTHS

8

CREATING ET: SCIENCE FICTION AS FUTURISTIC MYTH

I am simply a story-teller who happens to be a student of science. If a man writes the best that is in him, he cannot help some of his serious speculations appearing.

—*H. G. Wells, science fiction and science textbook author, interview with* Weekly Sun Literary Supplement *(1895)*

Science fiction does not simply borrow authority from science, but...it also shapes science in profound ways.

—*James A. Herrick,* Scientific Mythologies: How Science and Science Fiction Forge New Religious Beliefs *(2008), 261*

Part one of this book critiqued seven myths about the alleged perpetually warring relationship of science and Christianity. These myths did not arise coincidentally. In part two we will dig deeper, further exposing the sources, real and imagined, for the origin and perpetuation of these warfare myths. Science fiction has contributed imaginatively, and so did a 1980 TV series. Today's myth perpetuators often cite the work of past scientists to justify their warfare narrative. The harmony of science and Christianity, however, will become clear through a case study of the life and work of the brilliant astronomer Johannes Kepler.

This chapter will explore how clever earthly minds created ET. Science fiction became the genre within which a new kind of myth

emerged: futuristic techno-scientific myth. The AI-ET enlightenment myth is a child of the futuristic mythology of sci-fi.

Science fiction shapes science, especially speculative scientific
inquiry into the likelihood and cultural impact of the AI-Singularity
and the arrival of ET. Arthur C. Clarke, the influential science writer
and science fiction writer who prophesied, "Any sufficiently advanced
technology is indistinguishable from magic," cowrote the screenplay
for Stanley Kubrick's film *2001: A Space Odyssey* (1968), which depicts
the horror and euphoria of an encounter with techno-magical AI
and ET. The movie ends with the starchild, a human fetus float-
ing serenely in space. This appears to be the new humanity after ET
enlightenment. For those disenchanted with traditional religion,
2001 sketched a space-age faith. Clearly, Clarke's imagination has
shaped some of the mythic elements of the AI-Singularity and SETI.

In the 2017 film *Transformers: The Last Knight*, the fifth installment of the blockbuster film franchise, the hero, Texan inventor
Cade Yeager (Mark Wahlberg), introduces himself as a "magician"
to the lovely Oxford professor Viviane Wembly (Laura Haddock).
To justify this surprising professional identity, Yeager quotes Clarke's
famous line about advanced technology as magic. As it happens,
Wembly is a descendent of King Arthur's legendary magician, Merlin. The film explains Merlin's magic, and other extraordinary historical events, as episodic interventions of aliens and their conscious
robots. Here is just another reason AI-ET enlightenment enthusiasts
welcome the *Transformer* movies, which break down the traditional
dichotomy between living flesh and artificially smart machines.

Blockbuster films often synthesize historic myth and futuristic
myth. Numerous sci-fi and superhero films have become highly prof-
itable by this approach. Notable examples include the iterations of
Thor (2011, 2013, and 2017) and that character's role in the larger
Marvel Cinematic Universe, the world's top-grossing movie franchise. If the story disappoints, at least the visual effects and the common desire for more control of one's environment draw the crowds
back to their reclining theater seats and bottomless popcorn.

The early history of sci-fi displays the roots of science-themed
mythology. Whereas ancient myth typically focused on the *past* and
its subsequent human implications, the German genius Johannes
Kepler helped pioneer sci-fi as modern myth in which we explore

the *future* of scientific knowledge and technological control of nature and what this means for humanity. Sci-fi, Kepler's included, typically envisions life through alien eyes in ways that spur reconsideration of our own world.

Kepler's Sci-Fi Dream

Kepler framed his story as a dream in which a lunar inhabitant called a Daemon transports earthlings to the moon to tutor them in astronomy and geography as practiced in this alien world.[1] The Daemon is a character from Greco-Roman myth that Kepler used to represent his new approach to astronomy.[2] Kepler adapted ancient myth, including Platonic philosophical fable, to create a new techno-scientific tale aimed, in part, at Copernican persuasion.[3] He did this for the glory of God, who may have created different kinds of intelligent creatures in various cosmic locations.

The lunar Daemon begins his spaceflight speech: "Fifty thousand German miles up in the ether lies the island of Levania [the moon]. The road to it from here or from it to this earth is seldom open. When it is open, it is easy for our kind, but for transporting men it is assuredly most difficult and fraught with the greatest danger to life. We admit to this company nobody who is lethargic, fat, or tender."[4] Only physically fit astronauts will do! The Daemon explains the journey to his lunar home—mostly in natural, not supernatural, terms—anticipating some of the challenges of space travel:

> In every instance the take-off hits him as a severe shock, for he is hurled just as though he had been shot aloft by gunpowder to sail over mountains and seas. For this reason at the outset he must be lulled to sleep immediately with narcotics and opiates. His limbs must be arranged in such a way that his torso will not be torn away from his buttocks nor his head from his body, but the shock will be distributed among his individual limbs.[5]

That is only the beginning of spaceflight troubles. Kepler continued: "Then a new difficulty follows: extreme cold and impeded breathing. The cold is relieved by a power which we are born with;

the breathing, by applying damp sponges to the nostrils."[6] This use of narrative fiction to envision future empirical investigation and technological control is a typical feature of sci-fi.[7]

Guided by the Daemon, humans encounter strange lunar species. The land creatures are of "monstrous size" and undergo "very rapid" growth but have short lifespans—all appropriate for that particular environment, which includes extreme swings in temperature and water availability caused by a day-night cycle equivalent to an earthly month (Kepler correctly described most of the astronomical details).[8] All of this was tied to Kepler's biblical teleology: "Surely everything was destined for a definite use."[9] Imagined exotic alien life was constrained, as is customary in sci-fi, only by what seemed *not* to be scientifically *impossible*. This left more room for speculation than was appropriate for science proper.

As humans encounter Levania (the moon) with the help of its highly intelligent native Daemons, they experience a new home that seems to be just as motionless as any German town. When they look skyward while standing on the hemisphere of Levania that always faces Volva (earth), what do they enjoy seeing most? In Kepler's dream we learn that "the most beautiful of all the sights on Levania is the view of its Volva."[10] Levanians (lunarians) experience their Volva in some ways that approximate how earthlings see their moon, most notably a pattern of repeating phases—full, half, and various in-between shapes (crescent and gibbous) over the same period of time as an earthling "month." But now the jolting difference: "Their Volva remains fixed in place, then, as though it were attached to the heavens with a nail. Above it the other heavenly bodies, including the sun, move from east to west. There is never a night in which some of the fixed stars in the zodiac do not pass behind this Volva and emerge again on the other side."[11]

Note how the inhabitants on the Volva-facing side of Levania always see their Volva at the same angle above their horizon.[12] The earthly readers of this celestial dream recognize (with Kepler's help) that our North Star serves as a similar marker of latitude on our planet. Levanians have their own north star but also their Volva, both of which help them know where they are located on their spherical home.[13]

Kepler's dream of aliens and humans trading places helps us grasp two different "world pictures"—the cosmos as seen and articulated from their world and ours. He launched sci-fi aimed at scientific discovery and technological control for the glory of God. Kepler helped create a literary portal by which we can see the cosmos through alien eyes and thereby better grasp our own experience of the world.

Although Kepler did not invent all that sci-fi would become, his dream was a remarkable achievement in imaginative archetype: modern myth in the service of factual science and anticipated technology. Indeed, Kepler pursued this line of inquiry as seriously as he sought to account for the Star of Bethlehem that signaled the birth of Jesus. In episode 3 of his 1980 TV series *Cosmos*, astronomer and epic storyteller Carl Sagan (echoing numerous scholars) celebrated Kepler's imaginative achievement as "one of the first works of science fiction" that, in part, became scientific and technological fact. Of course, most sci-fi is unlikely to become fact, but why spoil the fun prematurely?

Kepler's sci-fi was driven by a futuristic scientific pedagogical aim. He proclaimed, "The purpose of my *Dream* is to use the example of the moon to build up an argument in favor of the motion of the earth, or rather to overcome objections taken from the universal opposition of mankind." Imagining existence in an alternative cosmic location would help free earthlings from an "ancient Ignorance" that "still struggles on in a tangle of so many knots tied tightly together through so many centuries," Kepler insisted.[14]

So much for Kepler. How did sci-fi become a dominant source of futuristic narratives for scientists and popular culture? H. G. Wells deserves much of the credit for this transformation. But his vision of humanity's cosmic future radically departed from Kepler's.

H. G. Wells's Cosmic Spirituality Without God

Science fiction, known until the 1930s primarily as "scientific romance," matured to a widely recognized literary genre with the appearance of H. G. Wells's (1866–1946) early novels, especially *The Time Machine* (1895), *The Island of Doctor Moreau* (1896), *The Invisible Man* (1897), and *The War of the Worlds* (1898). The epigraph to the last deserves special attention:

But who shall dwell in these Worlds if they be inhabited?... Are
we or they Lords of the World?... And how are all things made
for man?
    —Kepler (quoted in *The Anatomy of Melancholy*)[15]

The Kepler epigraph, read in its Wellsian context, gives the impression that this great astronomer questioned Judeo-Christian human exceptionalism on the authority of science and especially Copernican astronomy. To the contrary, Kepler celebrated the harmony of the Bible with his own monumental scientific work (see chapter 10). Wells, in contrast, in both his sci-fi and in his unremarkable scientific prose in popular magazines, science journals, and a few textbooks, imagined a war between science and Christianity.

For example, in *The War of the Worlds*, which chronicles a Martian invasion of earth, the narrator questions the idea that humans are made in God's image: "I felt the first inkling of a thing that presently grew quite clear in my mind, that oppressed me for many days, a sense of dethronement, a persuasion that I was no longer a master, but an animal among the animals, under the Martian heel."[16]

In an analysis of *The War of the Worlds*, scholar Robert Crossley zooms in on Wells's assault on human exceptionalism:

> In the history of speculation about Mars this is not the first articulation of the notion that Mars is a mirror and Martians are reflections of human beings, but it is the most daringly designed. That Wells used the motif of seeing ourselves in the Martians in the first novel to portray them as monsters rather than as human facsimiles is characteristic of what he later described as his impulse at the time to make "an assault on human self-satisfaction."[17]

Of the many worlds (planets and moons) considered plausibly habitable at the end of the nineteenth century, Mars was the best candidate for life, and the subject of intense scientific debate. In 1889 leading astronomer (and highly successful textbook author) Charles Young made this telling statement about Mars's habitability: "Probably there is no astronomical subject concerning which opposite opinions are so positively and even passionately held."[18] (The same is true today about the alleged anti-theistic impact of imminent ET contact.) As a science popularizer and sci-fi author, Wells sided passionately with those who believed there could be life on Mars, but he discouraged conceiving of Martians as physically similar to humans. Such anthropomorphism reflected "cosmogonies and religions invented by the childish conceit of primitive man," he insisted.[19]

Soon after publishing what he called his first "scientific romance," *The Time Machine*, Wells identified himself in a way that he is most remembered today: "I am simply a story-teller who happens to be a student of science." Wells also said in this 1895 interview, "If a man writes the best that is in him, he cannot help some of his serious speculations appearing." Finally, he affirmed the value of fiction for communicating science—especially exciting speculative science—to the public: "You cannot blame science for welcoming so popular an expression."[20]

Wells opened *The War of the Worlds* with this announcement: "No one would have believed in the last years of the nineteenth century that this world was being watched keenly and closely by intelligences greater than man's and yet as mortal as his own."[21] He then launched a payload of passionate speculation and romantic disregard for the actual history of ideas about extraterrestrial life:

> At most terrestrial men fancied there might be other men upon Mars, perhaps inferior to themselves and ready to welcome a missionary enterprise. Yet across the gulf of space, minds that are to our minds as ours are to those of the beasts that perish, intellects vast and cool and unsympathetic, regarded this earth with envious eyes, and slowly and surely drew their plans against us.... Yet so vain is man, and so blinded by his vanity, that no writer, up to the very end of the nineteenth century, expressed any idea that intelligent life might have developed there far, or indeed at all, beyond its earthly level.[22]

Wells's story falsely constructed its own originality. For many centuries astronomers and theologians had debated the likelihood and possible evil or good nature of extraterrestrial intelligent creatures—especially in many-worlds speculation, as we saw in chapter 4. Even the prospect of vastly superior ET had been considered. ET had been a common topic throughout the history of sci-fi since Kepler. ET lore morphed over time, but not in the sudden manner Wells suggested.[23]

One of the novelties in Wells's *War of the Worlds* was his Darwinian vision of ET as utterly different from any life-form on earth—explicable, so Wells imagined, by the radically unearthly environment of Mars and the contingencies of organic evolution on any planet (Wells had sat under the teaching of Darwin's leading disciple, Thomas Henry Huxley). Behold the Martians:

> They were, I now saw, the most unearthly creatures it is possible to conceive. They were huge round bodies—or, rather, heads—about four feet in diameter, each body having in front of it a face. This face had no nostrils—indeed, the Martians do not seem to have had any sense of smell, but it had a pair of very large dark-coloured eyes, and just beneath this a kind of fleshy beak. In the back of this head or body—I scarcely know how to

Figure 8.1: Artist Henrique Alvim Corrêa's rendering of the Martians in War of the Worlds *closely follows Wells's description. H. G. Wells,* La Guerre des Mondes, *trans. Henry-D. Davray (Jette-Bruxelles: L. Vandamme, 1906).*

speak of it—was the single tight tympanic surface, since known to be anatomically an ear, though it must have been almost useless in our dense air. In a group round the mouth were sixteen slender, almost whiplike tentacles, arranged in two bunches of eight each. These bunches have since been named rather aptly, by that distinguished anatomist, Professor Howes, the HANDS. Even as I saw these Martians for the first time they seemed to be endeavouring to raise themselves on these hands, but of course, with the increased weight of terrestrial conditions, this was impossible. There is reason to suppose that on Mars they may have progressed upon them with some facility.[24]

Wells's attack on human exceptionalism is best expressed in an exchange between an unpleasant Christian minister (who sounds like H. G.'s perception of his Christian mother, Sarah Wells) and the novel's narrator:

"What does it mean?" he [the minister] said. "What do these things mean?" I stared at him and made no answer.

He extended a thin white hand and spoke in almost a complaining tone. "Why are these things permitted? What sins have we done? The morning service was over, I was walking through the roads to clear my brain for the afternoon, and then—fire, earthquake, death! As if it were Sodom and Gomorrah! All our work undone, all the work. What are these Martians?"

"What are we?" I answered, clearing my throat.[25]

In his biography of Wells, Michael Sherborne observes, "It is highly characteristic of Wells that, even as he parodies religious mythology, he depends on it to give shape to his ideas."[26] A similar pattern is apparent in popular scientific literature and even in some astronomy textbooks. Although Wells reported that he decisively rejected Christianity as a boy after a hellish nightmare, his mother's Christian influence continued to haunt him for much of his life, Sherborne documents. When she died in 1905, Wells "photographed his mother's laid-out corpse from several angles in close-up," which Sherborne says indicated the "deep significance of the bereavement for him." But now "that Sarah would no longer be monitoring his exploits he would become much more openly militant as a champion of socialism and a rebel against sexual propriety."[27]

Herbert George Wells's "truest disciple," known principally by his Wellsian pseudonym George Orwell (1903–1950), turned against Wells shortly before death overtook the master. Sherborne chronicles the fight that erupted between the two during Hitler's world rampage: many agreed with Orwell that Wells's prophetic "storytelling gift... had been overwhelmed by an obsessive idealism" and "a reductive belief in scientific progress." Wells's letter responding to Orwell's accusations has been lost except for the words "I don't say that at all. Read my earlier works you shit."[28] Sherborne's conclusion is telling, especially because his biography is known for leaning pro-Wells: "This reference to 'earlier works' gives the game away. Wells's earlier fiction was broader in scope than Orwell allowed, yet Orwell was right to suggest that most of Wells's later fiction was unbalanced by a superficial vision of redemption, rationalized by often illegitimate appeals to science."[29]

C. S. Lewis would have agreed with much of this assessment of Wells. In the final book of his space trilogy, *That Hideous Strength* (1945), Wells is caricatured as Horace Jules, the "distinguished novelist and scientific populariser" who had become the (publicly prominent but institutionally powerless) director of the National Institute for Coordinated Experiments (NICE)—a pseudo-scientific agency aimed at controlling the world and destroying traditional religion.[30] Wells's novel *The Shape of Things to Come* (1933), framed as a dream (echoing Kepler) about a history textbook published in 2106, depicted favorably the rise of a world government that abolished the great world religions. Techno-scientific bureaucrats seized the moment to create a new naturalistic faith. "There was now to be one faith only in the world," Wells announced. A "new education of all" would facilitate this and also would bring world peace. Wells insisted that "not only the collective organization of the race but the moral making of the individual had to begin anew."[31] This superficial vision of redemption is also at work in the ET-AI enlightenment myth.

The closing scene of the movie *Things to Come* (1936), as Wells scripted it, enlarged this vision:

OSWALD CABAL: [Humanity] must go on, conquest beyond conquest. First, this little planet and its winds and ways. And then all the laws of mind and matter that restrain him. Then the planets about him, and, at last, out across immensity to the stars. And when he has conquered all the depths of space, and all the mysteries of time, still he will be beginning.

RAYMOND PASSWORTHY: But we're such little creatures. Poor humanity's so fragile, so weak. Little, little animals.

OSWALD CABAL: Little animals. And if we're no more than animals, we must snatch each little scrap of happiness, and live, and suffer, and pass, mattering no more than all the other animals do or have done. It is this, or that. All the universe or nothing. Which shall it be, Passworthy? Which shall it be?[32]

Sci-Fi and Today's Search for ET-AI Enlightenment

The alleged demotion of humanity at the hands of Copernicus and Darwin could not restrain Wells and his followers from imagining a techno-scientific exalted humanity.[33] In like manner, Arthur C. Clarke in *2001: A Space Odyssey* introduced an ET-assisted and AI-enhanced humanity. Unfortunately, some undesirable humans had to be destroyed along the way. ET's technology appears magical, as the film illustrates powerfully with psychedelic hippie-generation special effects. Shockingly, this occult-like futuristic expectation pervades our culture's current ET-AI enlightenment myth, as chapter 7 documented. Yesterday's sci-fi has become today's naturalistic ideology.

Many recent astronomers and SETI leaders, echoing the sci-fi of Wells and others, indulge in mythmaking that paradoxically demotes and exalts humanity—sometimes explicitly offering an ET-inspired spirituality without God. Such future-oriented myths build on the historical warfare myths about science and religion. They do so by claiming that we must lay aside past anti-science beliefs in the supernatural and replace them with a techno-scientific agenda for transforming humanity. Yet the envisioned future world will feel quite spiritual partly because of the magical wonders of scientific technology. We have seen how sci-fi provided the first reading of these subjects, thus shaping subsequent science and its associated anti-theistic myths.

In his space trilogy C. S. Lewis set out to reclaim sci-fi from this sort of Wellsian reconstruction of humanity—and along the way he anticipated some of the excesses of SETI and AI-Singularity ideology. Lewis reaffirmed the Christian roots of sci-fi that reach back to Kepler's pro-Copernican, pro-ET space-travel dream. Kepler imagined life on other worlds without constructing a Wellsian godless demotion-promotion for humanity. The cosmic vision of Kepler or Wells—which shall it be, Passworthy? Which shall it be?

In chapter 10 we will become better acquainted with Kepler as one of the founders of modern science. In contrast to the warfare myths, his life and work illustrate the compatibility of Christianity and science. But to better recognize Kepler as a myth buster, we first need to venture into the world of televised cosmological storytelling. There we will find the TV documentary expression of futuristic sci-fi mythology. Like Wells, celebrity scientists end up finding religion in the attempt to escape it. Often this means that they develop an unconscious theology in the process of denouncing theology.[34] There are, no doubt, many questionable TV preachers. The materialistic TV preachers that we will explore next do not beg for money; they want only your soul—naturalized and pulverized into stardust.

9

PREACHING ANTI-THEISM ON TV: *COSMOS*

Lord God, who is and who was and who is to come, the Almighty.
—*Revelation 1:8, ESV*

The cosmos is all that is, or ever was, or ever will be.
—*Carl Sagan, opening statement,* Cosmos: A Personal Voyage, *1980 TV series*

The message is simple yet majestic, in almost biblical fashion. While strolling toward a coastal cliff between Big Sir and Carmel, California, Carl Sagan opened his 1980 *Cosmos* TV series in praise of the cosmos: "The cosmos is all that is, or ever was, or ever will be."[1] The universe is the eternal self-existent one, the Almighty. No TV preacher has ever reached a larger audience than Sagan. His message? Science saves us from primitive religious ignorance and fear. The cosmos takes God's place. Was Sagan's belief materialism or pantheism? Your call. It was definitely anti-theism.

More than a half billion people have watched Sagan's thirteen cinematic sermons. Neil deGrasse Tyson's clever and expensive remake of the series in 2014—along with a 2019 follow-up—now passes Sagan's gospel to a new generation. The *Cosmos* original and its remake are an instance of anti-theistic apostolic succession. The *Star Trek* series and its *Next Generation* successor are another example with a similar message.

These priestly TV voices proclaim "the cosmic perspective,"[2] preaching this myth by blending science, science fiction, naturalistic theology, and historically inaccurate accounts that bash Christianity as a hindrance to scientific progress. The sci-fi created by H. G. Wells and his literary descendants provided the *Cosmos* storytellers with beguiling futuristic techno-scientific myths easily adaptable to TV documentary. In fact, Sagan had soaked himself in sci-fi and was a postwar UFO believer (he later traded UFOs for SETI).[3] In the 1950s his University of Chicago dorm room was buried under stacks of sci-fi magazines.[4] Sagan heavily influenced Tyson, and so the sci-fi roots of the two *Cosmos* series are formidable.

Sagan's Cosmic-Perspective Myth:
Science, Sci-Fi, and Naturalistic Theology

Sagan's opening *Cosmos* scene echoed H. G. Wells's movie *Things to Come* (1936). Remember that Wells had his character Oswald Cabal, with a dreamy upward gaze, announce our destiny: Humanity "must go on, conquest beyond conquest. First, this little planet...then all the laws of mind and matter that restrain him. Then the planets about him, and, at last, out across immensity to the stars." Sagan's monologue emulated this optimistic sci-fi vision: "Our contemplations of the cosmos stir us—there's a tingling in the spine, a catch in the voice, a faint sensation, as if a distant memory, of falling from a great height. We know we are approaching the grandest of mysteries."[5]

Sagan believed that the cosmos created us from stardust without supernatural guidance. When intelligence emerged and cultivated science, this meant that the universe had become self-conscious— the "grandest of mysteries." "We are a way for the cosmos to know itself," he passionately declared in this first episode, and in two other episodes in the series.[6] Intelligent life was the hero of the cosmic story—that's us, and the more advanced ETs out there. We are the universe awakened. Imagination makes this worldview possible. It is an *imaginative archetypal narrative*, which means that it is a myth in the anthropological sense.

In the opening scene, Sagan said:

The size and age of the cosmos are beyond ordinary human understanding. Lost somewhere between immensity and eternity is our tiny planetary home, the earth. For the first time we have the power to decide the fate of our planet and ourselves. This is a time of great danger, but our species is young, and curious, and brave. It shows much promise.... I believe our future depends powerfully on how well we understand this cosmos in which we float like a mote of dust in the morning sky.

Sagan delicately balanced apparent human insignificance in the face of cosmic immensity with bold confidence in our power to seize our own future—one that "shows much promise." This is right out of the sci-fi playbook of H. G. Wells. It's godless yet, in a poetic way, spiritual. C. S. Lewis called it Wellsianity.[7]

Mythology expert Gregory Schrempp explains the challenge for science popularizers like Sagan: "In urging us to foreswear the vision of ourselves as preordained pinnacle of the natural world, the great challenge to the popular science writer is to construct an attractive alternative—a compensatory vision."[8] Sagan's compensatory vision had a few simple components. First, "we are a way for the cosmos to know itself." This self-knowledge might help us save Mother Earth and her smart offspring. If so, then eventually we will join the enlightened alien civilizations that have successfully navigated the adolescent cliff on which we presently stand—peering over the edge of nuclear self-destruction. Tapping his skull for emphasis, Sagan invited viewers to climb aboard the "spaceship of the imagination" for fantastic journeys through space and time. We must ascend into the heavens to acquire the knowledge needed to return and save ourselves. As the TV series depicted it, the spaceship's interior had a vaulted, cathedral-like ceiling. Sagan was seated as high priest. This was science, sci-fi, and naturalistic theology mingled in a spectacular party.[9]

Cosmos traveled the lofty road of the 1968 film *2001: A Space Odyssey.* That space drama featured one of the cleverest fast-forward sequences in cinematic history. An ape-like creature had just learned to use a bone as a tool. Aliens inspired him through a mysterious black monolith—an artificially intelligent, gigantic, upright, hovering domino (without the white spots). The ape man's new bone tool, tossed to the wind triumphantly, became a spaceship in the year 2001, with the artificially intelligent H.A.L. 9000 computer onboard to run the mission. *Cosmos* episode one featured a similar fast forward. Sagan held a dandelion seed, waving in the ocean breeze. "Come with me," he said as he released the seed, a vehicle of life. A brilliant dandelion-shaped spaceship suddenly appeared in the blackness of space. Was Sagan imitating Jesus, who called some of his disciples from the Galilean seashore with "follow me"?[10]

Soon after the series finished airing on TV, political scientist David Paul Rebovich recognized Sagan's ET enlightenment myth. In a 1981 essay, Rebovich built up to this point by noting how *Cosmos* presents leading scientists as "forward-looking intellectuals who dared conventional belief, even in the face of public derision and persecution." The scientists who made the big discoveries were the "good guys" of history. Rebovich addressed Sagan's ET prophecy:

> This overly simple historical interpretation is carried to an embarrassing extreme when Sagan projects our future contact with advanced extraterrestrial beings. Since these beings will be superior scientists, we have nothing to fear. As long as mankind demonstrates its respect for knowledge, our contact with extraterrestrials will be marked by mutual respect and entail an interplanetary cultural vibrancy.[11]

Sagan, like Paul Davies (as we saw in chapter 7), believed that scientific and technological advancement automatically comes with moral superiority. That would make contact with ET a way to increase our chances of salvation. In 1985 Sagan explored this idea further in his sci-fi novel *Contact*, which in 1997 became a movie by the same name, starring Jodie Foster. The movie was released soon after Sagan's death. In both the novel and the movie, the hero scientist played out Sagan's answer to the question "Who Speaks for Earth?"—which was the title of the last episode of *Cosmos*. He meant, "Who will speak for us to ET?" Sagan's answer: scientists. They are the mediators between godlike ET creatures and ordinary humans. Sagan, if pressed, might have narrowed that down further. The most interdisciplinary and reputable scientific communicators who know well "the cosmic perspective" should speak for earth—scientists like, well, Sagan himself.[12]

In *Cosmos*, Sagan blended futuristic, sci-fi-inspired myth with traditional, past-oriented religious myth that he had reformulated in a
materialistic or pantheistic manner. His creation myth proclaimed us
to be the product of a material process with no deity involved. And yet,
in pantheistic style, the cosmos as a whole became a God substitute.

Sagan's passion for sci-fi was reflected in the futuristic mythology
of his *Cosmos* series. He urged us to look for that blessed hope and
glorious appearing, when the aliens will make contact and enlighten
us. Meanwhile, we should attempt to save ourselves with better technology and a shared scientific ethos of open exchange of ideas among
all people. This was almost as naive about human foibles as H. G.
Wells's later sci-fi.

Southern novelist Walker Percy answered Sagan with the book
Lost in the Cosmos. Percy tipped his hat to the cleverness of Sagan's
Cosmos while identifying its "unmalicious, even innocent, scientism,
the likes of which I have not encountered since the standard bull
sessions of high school and college—up to but not past the sopho-
more year."[13] In response to Sagan's search for ET as an antidote to
cosmic loneliness, Percy asked, "Why is Carl Sagan so lonely?" Percy
ventured a response: "Sagan is lonely because, once everything in
the Cosmos, including man, is reduced to the sphere of immanence,
matter in interaction, there is no one left to talk to except other tran-
scending intelligences from other worlds."[14]

Sagan's Imagined War: Science vs. Christianity

Consider another kind of myth in Sagan's *Cosmos* series: telling false
stories about history. He constructs some woefully inaccurate accounts
of the history of science and Christianity. These historic myths help
support Sagan's atheist-pantheist myth of "the cosmic perspective,"
which assumes that smart people don't believe in the God of theism.
In chapter 2 we saw how he perpetuated the myth of the Dark Ages.
In his TV series he marked the beginning of the Dark Ages with a
misconceived event—the death of the philosopher Hypatia in 415.

Sagan described Hypatia as "a mathematician, astronomer, physicist, and head of the school of Neoplatonic philosophy in Alexandria." He told his viewers:

The growing Christian Church was consolidating its power and attempting to eradicate pagan influence and culture.... Cyril, the bishop of Alexandria, despised her in part because of her close friendship with a Roman governor but also because she was a symbol of learning and science, which were largely identified by the early Church with paganism.

In great personal danger, Hypatia continued to teach and to publish until, in the year 415 AD, on her way to work she was set upon by a fanatical mob of Cyril's followers. They dragged her from her chariot, tore off her clothes, and flayed her flesh from her bones with abalone shells.

Her remains were burned, her works obliterated, her name forgotten.

Cyril was made a saint.[15]

Others have amplified Sagan's error. Historian of science Michael Shank reports, "In 2009, Alejandro Amenábar's movie *Agoura* used Sagan's stunningly anachronistic history of science to make the 5th-century murder of Hypatia in Alexandria the beginning of the Dark Ages."[16] By "anachronistic" Shank means inappropriately imposing present ways of thinking onto the past, thereby distorting history.

Was Hypatia, as Sagan taught TV viewers, a martyr for science? Was Christian teaching to blame for the rioting crowd that murdered her? In fact, her vibrant philosophical school and civic work had little to do with science. Furthermore, her death was politically motivated and highly circumstantial.

Historian Edward J. Watts's excellent book about Hypatia corrects most of Sagan's false account. Watts reports that Hypatia succeeded her father, Theon, as head of a school that had long emphasized mathematics, including astronomy. Sadly, the astronomy of Theon and Hypatia was in a less vigorous condition compared to the golden years of the great Alexandrian astronomer Ptolemy two centuries earlier. In the 380s Hypatia devoted herself and her pupils to a Neoplatonic philosophy that emphasized nurturing a well-ordered soul to better commune with divinity. The material world had corrupted the human soul, she thought. Her school aimed at developing virtues that would direct the soul back to its immaterial divine source. When Hypatia began teaching philosophy, Alexandria in Egypt had recently become a Christian-majority city. So most of her students were Christians. She successfully engaged both pagan and Christian students in philosophical education by keeping the nature of divinity sufficiently vague and by shunning pagan ritual as a cheap substitute for Neoplatonic philosophical spirituality.[17]

Hypatia's world of peaceful pagan-Christian dialogue began to disintegrate in the 390s. Theophilus, the new archbishop of Alexandria, led a parade that mocked pagan rituals. Pagan mobs turned violent in protest. They took Christian prisoners and retreated to the pagan temple on Serapeum hill, which stood as the Alexandrian equivalent of Athens's Acropolis. Finally, the Christian emperor Theodosius granted blanket amnesty to the pagan rioters and they dispersed. Then a Christian crowd took over the temple and destroyed the gigantic statue of Serapis that was the principal icon of this cult.[18]

When archbishop Theophilus died in 412, another Christian faction contested the appointment of his successor, Cyril. More violence erupted. Cyril pushed back at this faction and its allies, including Jews. Orestes, the Roman governor of the province of Egypt (and a Christian), punished a strong Cyril supporter accused of provoking civil unrest. A long series of imprudent moves by both Orestes and Cyril escalated to a critical point. Orestes called on Hypatia as a neutral party to build a coalition that might stabilize the city and oppose Cyril.

Rumors about Hypatia's bewitching Orestes triggered a response. A group of rowdy people gathered in March 415 at the instigation of a Cyril supporter named Peter. His opportunistic crowd probably consisted of seaport workers and sailors during a month usually short on that sort of employment. So this crowd probably included desperate people, especially non-Alexandrian citizens with no access to government grain assistance. The plan almost certainly was to intimidate, not kill, Hypatia. That was typical political action in late antiquity, regardless of the affiliation of the activists. It usually went like this: Shout and scream outside the walls of an influential person's home, and then disperse. This mob, however, happened to encounter Hypatia in public. With passions flaring, they shredded her body using broken roof tiles—tools of spontaneous violence, not premeditated murder. Then they cremated her mangled remains beyond the city limits, in accordance with traditional purification symbolism that had episodically rid the city of persecuted Christians and other outcasts during the previous two centuries.[19]

Virtually no early accounts of the event claimed that Cyril ordered an attack on Hypatia, though there was wide agreement that he helped create the climate that triggered it. The church historian Socrates Scholasticus (ca. 380–ca. 450) wrote: "This affair brought no slight opprobrium upon Cyril and the whole Alexandrian church. Indeed nothing is farther from the spirit of Christianity than murders, fights, and similar things."[20] Christians almost universally condemned this event as senseless murder. Political rivalry and personal imprudence, not Christian teaching, was to blame.

Nor was Hypatia a martyr for science. She taught a little astronomy, but only as a way to prepare the mind for what she thought was really important: purification and communion with divinity by rejecting the cares of this material world (even sex).[21] To the wider pagan and Christian world, even within Alexandria, Hypatia was (in Watts's words) "an accomplished philosopher whose research had made meaningful contributions to the intellectual life of the empire and whose public activities had shaped the political life of her city."[22]

Sagan claimed that Cyril despised Hypatia "in part because of her close friendship with a Roman governor but also because she was a symbol of learning and science, which were largely identified by the early Church with paganism." He got the first part right, but the second claim is a howler.

Hypatia's story is tragic but largely irrelevant to the history of science. Yet Sagan associated her murder with the destruction of the Alexandrian Royal Library and its science teaching and research activity. This also was allegedly the doing of a Christian mob. He called Hypatia "the last scientist to work in" the Alexandrian Royal Library, the world's leading library and scientific research facility. Sagan falsely asserted that the last remains of the library "were destroyed within a year of Hypatia's death." With even less historical justification, he mourned: "It's as if an entire civilization had undergone a sort of self-inflicted radical brain surgery so that most of its memories, discoveries, ideas, and passions were irrevocably wiped out. The loss was incalculable."

Here was the reality: The Royal Library had formed part of the Alexandrian Museum, but the entire institution of research, teaching, and book preservation had drastically dwindled before Hypatia was born, and before Christians formed a majority in Alexandria. The last reference to this original museum-library institution of research and teaching dates from the 260s. It was probably destroyed before 300.[23] Even before that, the quality of its work had diminished owing to imperial appointments guided by political rather than academic criteria.[24]

The museum appears to have been reestablished on a different site in the fourth century. Very little is known about the museum in this, its final phase, of the fourth and fifth centuries. Its last member, Horapollon, may have joined it after converting to Christianity.[25] Eventually Christians abandoned the old, failing institutions of learning and created new ones. Monasteries and cathedral schools preserved literature and extended learning. Finally, medieval Europe created universities that have continued as beacons of science and scholarship to our day.

Sagan's Alexandrian sermon in his final episode ended with a call to come forward in dedication: "History is full of people who, out of fear or ignorance or the lust for power, have destroyed treasures of immeasurable value which truly belong to all of us. We must not let it happen again." Agreed, and this includes not letting Sagan destroy the treasures found in the history of science and Christianity, with its predominant trait of harmony rather than conflict. The cultural resources of Christianity that have helped nurture the growth of science should be shared freely with all, even with those practitioners and supporters of science who no longer operate from a Christian worldview.

Team Tyson Re-creates *Cosmos* Featuring Bruno Superstar

Carl Sagan littered his TV series with yet other misconceptions about historic relations between science and Christianity, including myths about Bruno and Galileo debunked earlier in this book. Neil deGrasse Tyson's 2014 *Cosmos* treats history with a similar disregard for accuracy.

The lead writers for *Cosmos* 2014 had both helped Sagan write the 1980 series. One was Ann Druyan, who became Sagan's wife in

1981—and who received the 2004 Richard Dawkins Award from the Atheist Alliance of America.[26] The other writer was astrophysicist Steven Soter of New York University and the American Museum of Natural History. So we would expect to see significant continuity between the original *Cosmos* and its sequel. We do. But none of the writers for either series were historians of science. And it shows.

Cosmos 2014 begins with Sagan's haunting voice hailing the cosmos as "all that is, or ever was, or ever will be." Viewers not inclined to worship the universe will still be eager to climb aboard the CGI-upgraded "spaceship of the imagination" for exhilarating and educational journeys through space and time. Like its predecessor, much of the 2014 *Cosmos* offers good science education. Tyson is a superb communicator and projects a persona that puts everyone at ease.

But the myths appear early on. In fact, in episode one, in Tyson's very first sentence about the history of science, we hear him perpetuating what I identified in chapter 1 as the "big myth":

> Back in 1599, everyone knew that the Sun, planets, and stars were just lights in the sky that revolved around the earth, and that we were the center of a little universe, a universe made for us. There was only one man on the whole planet who envisioned an infinitely grander cosmos. And how was he spending New Year's Eve of the year 1600? Why, in prison, of course.[27]

Especially since the time of Ptolemy of Alexandria in the second century AD, educated people in the Western tradition, including Hypatia and her predominantly Christian students, have had a vivid picture of a huge universe. This knowledge persisted through the European Middle Ages and on up through the year 1600, and beyond. Tyson perpetuates the myth of a widely accepted tiny cosmos to advance still another myth debunked in this book: that Giordano Bruno became a martyr for science. (I use "Tyson" here as shorthand for Team Tyson—especially the writers of the show, Soter and Druyan. Sagan had much more control over the entire 1980 script.)[28]

For it's Bruno who, in Tyson's telling, is "the one man on the whole planet who envisioned an infinitely grander cosmos." Strangely, Tyson barely mentions Copernicus, and then only to say that Bruno exceeded his predecessor's excellence. Here is the *only* thing that

Tyson, in all thirteen episodes, says of the great astronomer Nicolaus Copernicus: "Copernicus made a radical proposal. The Earth was not the center. It was just one of the planets, and, like them, it revolved around the Sun. . . . But for one man, Copernicus didn't go far enough. His name was Giordano Bruno, and he was a natural-born rebel."

How did this rebel exceed the work of Copernicus? "He longed to bust out of that cramped little universe. Even as a young Dominican monk in Naples, he was a misfit. This was a time when there was no freedom of thought in Italy. But Bruno hungered to know everything about God's creation." Even though Copernicus had to greatly enlarge our vision of an already huge cosmos to explain away the lack of observed stellar parallax, Tyson still calls this Copernican universe too "little" for Bruno, even "cramped." Also at work here is the impression of big minds overcoming lesser ones.

This celebration of Bruno comes on the heels of Tyson's advertisement for multiverse theory (an infinite number of universes) as a successor to the well-established account of a single cosmic beginning at the Big Bang. The rhetorical device of *big think* versus *small think* is serving Tyson's agenda here as well. But greater physical size does not necessarily correlate with more significance. Tyson's rhetorical fog also distracts from the fact that the multiverse is hugely speculative. It offers only a vague mathematical impression of support to what is otherwise philosophical extravagance. Of course, God could have made other universes. Some Catholic officials insisted upon it in 1277, as we saw in chapter 4.

What did Bruno do next in Tyson's story? He read the ancient pagan Lucretius's thought experiment that suggested an infinite cosmos. "This made perfect sense to Bruno. The God he worshiped was infinite. So how, he reasoned, could creation be anything less?" Tyson leaves out the inconvenient fact that Nicholas of Cusa had argued for an infinite universe a century before Bruno. Mentioning this great cardinal of the Catholic Church as one of the main sources of Bruno's ideas would have diminished the warfare story.

Astrophysicist Steven Soter, the main writer of the *Cosmos* 2014 Bruno story, defended the account in dialogue with *Discover* magazine's Corey Powell, who studied history and science at Harvard. Powell won this blog debate with Soter by pointing to more evidence against Bruno as the "lone wolf" discoverer of some things now known to be true. For example, Powell noted that "even the idea that other stars are suns," attributed solely to Bruno in Soter's *Cosmos* cartoon, "has possible precedence a century earlier in the work of Nicolas of Cusa." He cited Nicholas of Cusa's reference to "the sun, or another star" to make that point. More important, Powell noted how *Cosmos* wrongly "depicts Bruno's infinite cosmology as a physical theory of the universe." He reported that Bruno, in a few of his books, "imagines all planets and stars having souls" and "uses his cosmology as a tool for advancing an animist or Pandeist theology." Pandeism is a variation of pantheism in which the creator becomes the cosmos. Powell concluded that Bruno's overall cosmology "was not a correct scientific idea, nor was it even a [correct] 'guess' as *Cosmos* asserts. It was a religious and philosophical statement.... That is why I say that religion, not science, caused Bruno's deadly clash with the Church." Although the Bruno story is a little more complicated than this, Powell was much closer to the truth than *Cosmos* 2014. Near the end of this blog, Powell, perhaps speaking for *Discover* as a popular science venue, said, "I greatly admire the entire *Cosmos* project, which is why I am being so critical here."[29]

Continuing Bruno's story, *Cosmos* tells viewers that the Dominican friar "assumed that other lovers of God would naturally embrace this grander and more glorious view of Creation." Bruno was wrong. "Heretic, infidel," critics yelled at him. "Your God is too small," Bruno replied in the core mythical nugget of Tyson's saga. Remember that in real history the Catholic Church had officially declared none of these ideas heretical.

At the end of the Bruno story comes what Tyson calls Bruno's "martyrdom" for the freedom to challenge Christianity and to embrace an infinite cosmos: "Ten years after Bruno's martyrdom, Galileo first looked through a telescope, realizing that Bruno had been right all along." Team Tyson just made up this bit about Galileo's confirming an infinite universe. His telescopic discoveries did nothing of the sort. Sure, Galileo saw many more stars than humans had previously estimated. Even so, Galileo, Kepler, and most other scientists realized that these telescopic discoveries in no way confirmed an infinite universe.

The Silent, Animated Passion of the Bruno Christ

The most powerful visual rhetoric in the 2014 *Cosmos* comes as an animated sequence that runs from Bruno's condemnation to his death. Dialogue and narration are absent. The animation and music deliver an ethereal messianic message.[30] It is the passion of the Christ.

Bruno, awaiting execution in his cell, peers longingly through a barred prison window. He steps back, shackled wrists clanking the adjoining chain. The camera zooms in on a dreary, unshaven face, sad eyes closed in resignation. Suddenly, with his mind's eye, Bruno sees his cosmic destiny. The defeated, bearded face is transfigured into a clean-shaven man at peace with the cosmos. Sunlight beaming through the prison window makes his body glow against the dingy stone walls of his Catholic confinement. He spreads his arms wide and levitates upward as the prison walls fall outward, revealing an infinite cosmos into which he ascends. He rockets his way between sun and moon. The shadows in the palms of his creased, outstretched hands are distinctly exaggerated at two points. A pair of nail holes suggest themselves. The ancient messianic imagery intensifies. One leg is cocked in traditional crucifix stance. Glorious music has replaced the earlier ominous tones. Bruno does a 360-degree spin as he jets past planets near an alien sun. The camera zooms in on his triumphal face, hair tossed freely by refreshing cosmic winds.

Suddenly, the crucified and risen cosmic Bruno fades back to a man of sorrow, riding on a donkey. The crowd jeers with raised fists. Two crosses on poles and an elevated battle-ax pierce the early morning sky as authorities escort Bruno to his place of execution. The tongue-tied Bruno is placed on the altar. Before the fire is lit, a crucifix on a pole is thrust in his face. The right leg of this crucified Jesus is bent at the same angle as the right leg of the Bruno-Christ in his earlier imagined death and resurrection. Bruno turns his scowling face away in rejection of the Catholic icon. The altar is engulfed in flames, and the divine sparks of Bruno flicker into the heavens to rejoin the stardust from which they came. The show then fast-forwards to Tyson in his spaceship declaring that "after Bruno's martyrdom" Galileo looked through a telescope, "realizing that Bruno had been right all along."

Bruno is a martyr for science. The message couldn't be more clear. But it's false. As a transition to the next segment of episode one,

Tyson admits: "Bruno was no scientist. His vision of the cosmos was a lucky guess, because he had no evidence to support it." But this comes too late. The earlier myth already has been lodged in the hearts of viewers: Bruno defied the Catholic Church in a heroic quest to embrace a scientific-looking infinite universe. Had Tyson presented an accurate history of Bruno, then his story would have lost its power to create reverence for an infinite cosmos or multiverse. For Tyson, the multiverse is God, and Bruno is its martyred prophet.

Justifying Anti-theistic Myths and Erroneous History for a Greater Good?

In the rebooted *Cosmos* series, Tyson's narrative of warfare between science and religion is pervasive. He imposes it on many other great thinkers, ranging from the ancient philosopher Thales of Miletus to the great physicist Isaac Newton.[31] Based on Tyson's past statements, his cartoonish history was to be expected. When Bill Moyers asked him whether faith and reason are compatible, Tyson replied, "I don't think they're reconcilable." Later he confessed why: "God is an ever-receding pocket of scientific ignorance."[32] Tyson told Moyers that science makes it increasingly more difficult to believe in theistic religion, as new discoveries replace God explanations with materialistic ones. But John Lennox, emeritus University of Oxford mathematics professor, has debunked this notion.[33]

At least one historian of science thinks that, for the sake of a greater good, it might be permissible for *Cosmos* 2014 to broadcast false history of science. Joseph D. Martin, historian and philosopher of science and technology at the University of Cambridge, wrote that he agreed with many of the criticisms of *Cosmos*'s representations of history, but he added, "*Cosmos* is a fantastic artifact of scientific myth-making and as such provides a superb teaching tool when paired with more responsible historical presentations."[34] Sadly, however, very few teachers have the necessary background in the history of science to responsibly critique the web of myths in *Cosmos*.[35] Martin continued:

> If we [grant] *Cosmos* the artistic license to lie, the question is then whether it [is] doing so in service of a greater truth and if

so, what is it? And what does it mean for us if it turns out that *Cosmos* and the history community are simply going after different truths? For the record, I myself am still very much on the fence about this issue, but if I were tasked with mounting a defence of *Cosmos* as it stands, one of the things I'd say is that the stakes of scientific authority are very high right now, especially in the United States. Perhaps the greater truth here is that we do need to promote greater public trust in science if we are going to tackle some of the frankly quite terrifying challenges ahead and maybe a touch of taradiddle in that direction isn't the worst thing.[36]

Taradiddle is a fancy word for a small and presumably harmless lie. What happened to the gold standard of striving after truth professionally, popularly, politically, and personally? Many of my fellow historians of science and philosophers of science still strive with me for this high standard. Martin's fence-straddling moral ambiguity is disturbing, perhaps even terrifying. Does anyone on Team Tyson think like Martin?

At least one does. The executive producer of *Cosmos* 2014 says that he has spent most of his professional life creating myths for the greater truth of atheism. His name is Brannon Braga. Speaking at the 2006 International Atheist Conference, he celebrated his part in creating "atheistic mythology" in more than 150 episodes of *Star Trek: Next Generation*. He summed up his mission—which violates the original *Star Trek* "prime directive" of not altering native culture—as showing that "religion sucks," "isn't science great," and finally "how the hell do we get the other 95 percent of the population to come to their senses?" These are remarkable confessions. As we saw in chapter 8, Kepler helped establish sci-fi as a way to promote very different ideas: "God rules the cosmos," "isn't science great," and finally "how for heaven's sake do we get the other 99.9 percent of the population to come to their senses so they can embrace Copernican astronomy?"

According to Braga, teaching atheistic myth is the work of sci-fi films *and* TV documentaries like *Cosmos*.[37] Indeed, he said that *Cosmos* 2014 was designed to combat "dark forces of irrational thinking." He emphasized: "Religion doesn't own awe and mystery. Science does it better."[38] But as we have seen, rendering Christianity as the historical enemy of science is itself an exercise in unreasonable and reckless historiography. Myth, not science, recognizes the cosmos as "all that is, or ever was, or ever will be." Sagan knew this statement would inspire awe because it imitated the biblical description of God. No doubt, Braga and his team of like-minded creators were delighted to rerun this mythical mantra at the beginning of *Cosmos* 2014. It served well the greater good of anti-theism.

Preaching "The Cosmic Perspective"

Cosmos 1980 and 2014 preach "the cosmic perspective" as an imaginative archetypal narrative. It takes a multigenerational team of diverse talent to keep a myth like this living perpetually. Carl Sagan, Ann Druyan, Steven Soter, Brannon Braga, Neil deGrasse Tyson, and others have formed such a team for 1980 and 2014. And *Cosmos* 2019 is coming.

So far, according to *Cosmos*, the superstars of human history have been Hypatia and Bruno. These characters dominate a fable of conflict between science and Christianity that is part of the cosmic perspective. The cosmic perspective, so its creators say, will one day wean us from all terrestrial religions, especially Christianity. Heroic warfare stories like these, no matter how badly at odds with historical evidence, are needed to hasten the universal acceptance of the cosmic perspective. But these warfare narratives are mostly myths in the ordinary sense of "false stories."

A completely cosmic perspective as Sagan and Tyson envision it must also be formulated as a myth in the anthropological sense. So what are the imaginative archetypal components of this sophisticated myth? First, the story tells us what is real: no heaven, no hell—just the cosmos, humans, and ET. Enlightened ET has found salvation through science and technology—no God required. So when ET arrives, humans will embrace this cosmic perspective with a confidence approaching certainty, so prophesy "the cosmic perspective" mythmakers. This is a grand vision of a theistically godless future.

The myth of the cosmic perspective blends science, sci-fi, naturalistic theology, and historically inaccurate accounts that bash Christianity as a hindrance to scientific progress. Sagan and Tyson think that the cosmos created us without a transcendent God. Once we all get the full story, uniting intelligences across galaxies, none of us will believe in a theistic God anymore—especially one that takes on *human* flesh, dies for *human* sins, and is raised in triumph over sin and death. That God, Bruno declared, is "too small" for the cosmic perspective.

The great astronomer Johannes Kepler thought otherwise. Why listen to him? Because he had something to say about *every* component of the cosmic-perspective myth: science, sci-fi, theology, and the history of science in relation to Christianity. He helped create sci-fi as futuristic storytelling for the glory of the God, and the advancement of Copernican astronomy. He also discovered three laws of planetary motion while under the influence of a perspective radically different from "the cosmic perspective" myth.

What was Kepler's perspective on the world? How did it help him discover many truths about the cosmos? Why is all of this a gigantic myth buster? That is the story of the next chapter.

10

KEPLER, DEVOUT SCIENTIST

[Kepler was] not a mystic, as is often claimed, but rather a man of his age, devout and rational at the same time.

—*Bruce Stephenson, The Music of the Heavens: Kepler's Harmonic Astronomy (1994), 15*

I am eager to publish (my observations) soon, not in my interest, dear teacher.... I strive to publish them in God's honor who wishes to be recognized from the book of nature.... I had the intention of becoming a theologian. For a long time I was restless: but now see how God is, by my endeavors, also glorified in astronomy.

—*Johannes Kepler, letter to Michael Maestlin, October 3, 1595*

God wanted us to recognize them [i.e., mathematical natural laws] by creating us after his own image so that we could share in his own thoughts.

—*Kepler, letter to Herwart von Hohenburg, April 9/10, 1599*

Geometry is unique and eternal, and it shines in the mind of God. The share of it which has been granted to man is one of the reasons why he is the image of God.

—*Kepler, letter to Galileo, 1610*

Johannes Kepler's life and work marvelously exemplify the support that Christianity and science have provided each other. This case study offers a high-profile counterexample to the thesis of endless warfare between science and faith. Kepler is an especially appropriate case because he was the most prominent scientist featured in Carl Sagan's *Cosmos*. Sagan managed to avoid acknowledging the multitudinous ways in which Kepler's work embodied the harmony of science and Christianity.

Disentangling Sagan's Kepler from the Historical Kepler

Here is an example of how Sagan misrepresented Kepler's Christian religious experience. He claimed that the young Kepler had "withdrawn into the world of his own thoughts, which were often concerned with his imagined unworthiness in the eyes of God. He despaired of ever attaining salvation."[1] Sagan implied that Kepler always felt this way. This sounds more like Sagan reflecting of his own conflicted religious experience growing up Jewish in Brooklyn.[2] His father was a bacon-eating Reform (liberal) Jew and his mother was a kosher Conservative Jew. Carl argued vehemently with his mother about God when he rejected theism as a teen.[3]

Although the exact chronology is unknown, for most of Kepler's life, including up through its end, he clearly understood his salvation to be a free gift from God and not something to be attained by human effort. Kepler perceived his own inherent unworthiness before a holy God as an objective fact, not a mere subjective experience. But he also claimed to have received the overwhelming love and forgiveness of God through Jesus Christ, which made him worthy to live joyfully in God's presence. He especially felt God's pleasure when doing astronomy. Shortly before his death, when asked about the basis of his salvation, Kepler responded that he was saved "solely by the merit of our savior Jesus Christ."[4]

As the epigraphs indicate, Kepler was a devout Christian who believed that the Bible and the "book of nature" were fully compatible and mutually supportive. He recognized them both as God's revelation. He studied both intensely. In fact, he almost finished a doctoral degree in theology before he turned to a career in mathematics and astronomy. Kepler believed that mathematical ideas exist eternally in the divine mind and that God freely selected some of these principles to govern his creation. Because God created humans in his image, we have the intelligence needed to discover those natural laws, and in so doing, Kepler announced, we "share in his own thoughts." The human mind emulates God's thoughts in ways that reveal the deep structure of the cosmos. Thus God is "glorified in astronomy," Kepler concluded.

He emphasized these points in his multivolume *Epitome of Copernican Astronomy* (1618–21), the world's first sun-centered text for astronomy education. In it he wrote that he considered it "a right, yes a duty, to search in cautious manner for the numbers, sizes, and weights, the norms for everything He has created."[5] Kepler focused on the way in which the human mind resembles God's mind, so that we can think like him:

> For He Himself has let man take part in the knowledge of these things and thus not in a small measure has set up His image in man. Since He recognized as very good this image which He made, He will so much more readily recognize our efforts with the light of this image also to push into the light of knowledge the utilization of the numbers, weights and sizes which He marked out at creation. For these secrets are not of the kind whose research should be forbidden; rather they are set before our eyes like a mirror so that by examining them we observe to some extent the goodness and wisdom of the Creator.[6]

In the dedication to *Epitome of Copernican Astronomy*, Kepler identified himself as a "priest of God, the creator of the book of nature." He commented that, as a Copernican advocate in the face of opposing "public prejudice," he wrote his textbook as a "hymn for God the Creator."[7] Contrary to Sagan's attribution of Kepler's "fear" of God, Kepler's joy before God, even in controversial subjects, permeated his astronomical works.

In *Epitome*, Kepler invited his student readers to consider how his new astronomy contributed greatly to "illustrating the glory of the fabric of the world, and of God the Architect." He added that "all philosophers, whether Greek or Latin, and all the poets too, recognize a divine ravishment in investigating the works of God."[8]

Three Laws of Planetary Motion:
Kepler Shares God's Thoughts

Centuries of astronomy textbooks have celebrated "Kepler's three laws of planetary motion," for which he is best remembered. Kepler would object to this designation. They are God's laws, not mine, he would insist. The women's education pioneer Emma Willard (1787–1870) recognized this point in her 1860 astronomy textbook, when she wrote, "They are laws of the Almighty, discovered by him."[9]

Kepler announced the discovery of two of his three laws of planetary motion in his 1609 book, *Astronomia nova* (New Astronomy). A series of unexpected events created the opportunity for Kepler to write this book. When Catholic authorities forced the intensely Protestant, and mostly Lutheran,[10] Kepler from his home and teaching position in 1599, he sought refuge with the world's greatest astronomer, who had just moved from Denmark to Prague. That man, Tycho Brahe (traditionally identified by his first name), had amassed a large collection of precise planetary observations that were in need of further analysis. Because Kepler was gifted mathematically, Tycho hired him. Soon after, the great Danish astronomer died, and the Holy Roman Emperor, Rudolf II, chose Kepler to replace him. Kepler gained full access to Tycho's trove of astronomical observations.[11]

Initially assuming circular planetary motion, as Tycho and Copernicus had done, Kepler found that Mars in theory deviated from Tycho's observed Mars by eight minutes of arc, which is about one-eighth of one degree (a tiny discrepancy in angular measurement). Perhaps God had selected some other geometrical closed curved figure besides the circle to govern planetary motion. This hypothesis developed, in part, because of Kepler's Christian belief that "geometry is unique and eternal, and it shines in the mind of God."[12] So Kepler toyed with variously shaped oval orbits, each with its own beautiful mathematical traits. In 1605 Kepler discovered that only the ellipse gave an accurate solution that also seemed to make physical-causal sense. He concluded that Mars travels in an elliptical orbit around the sun, and that the sun is the main physical cause of Mars's orbital motion.

Without Tycho's precise observational data, and Kepler's strong commitment to finding theories that closely match the best available evidence, many of Kepler's astronomical discoveries would have been impossible. Accordingly, in *Astronomia nova* Kepler thanked God for Tycho's data:

> Divine Providence granted us such a diligent observer in Tycho Brahe that his observations convicted the Ptolemaic calculation of an error of 8 minutes [of angular measurement]; it is only right that we should accept God's gift with a grateful mind.... Because these 8 minutes could not be ignored, they have led to a total reformation of astronomy.[13]

With his first law, Kepler repudiated a two-thousand-year-old tradition, established by Plato and perpetuated even by Copernicus and Tycho, of explaining the apparent motion of each planet through complicated *combinations* of *circular* motions. Kepler invoked a *single* geometric curve per planet. He called this physical path of a planet an "orbit."[14] He also coined the term "focus" to refer to each of the two defining points of an ellipse.[15] Like Shakespeare, Kepler left a trail of newly minted words behind his innovative work.

Kepler's second law specified that a planet moves in its elliptical orbit in such a way that the line joining the planet to the sun sweeps out equal areas during equal time intervals.[16] This meant that the time it takes for a planet to move a certain portion of its orbit is measured by the area swept out by the same joining line. The law also showed how a planet speeds up as it approaches the sun and slows down during the part of its orbit farther from the sun. Kepler concluded that the sun's force has a greater effect on a planet when it is closer to the sun.

Kepler announced the third law of planetary motion in his book *Harmony of the Universe* (1619). Perhaps the heavens display a harmony similar to the mathematical rules of musical harmony, he speculated. Eventually Kepler found what he was looking for: what textbooks often call the "harmonic law." Planets rhythmically dance to God's harmonious music.

The time it takes for a planet to orbit the sun once is called its period. The amount of time for this period depends on the average distance of a planet from the sun. Planets that are closer to the sun take less time to orbit the sun than planets that are farther away. Although Copernicus knew this general relation, Kepler discovered an elegant mathematical rule that expresses a constant relation between each planet's period and its average distance from the sun. The harmonic law states that the ratio of the period squared, divided by the distance cubed, is the same for all planets. Beautiful!

So the periods and distances of all the planets are harmoniously related in a musical order composed by the creator. Kepler was the first to hear this cosmic music. Although many of Kepler's harmonic speculations have been discarded, this law has endured subsequent testing: $Period^2/Distance^3$ is the same for every planet.

Kepler's Judeo-Christian Rationale for Discovering Natural Laws

Before discovering the three laws of planetary motion, Kepler worked out a Judeo-Christian rationale for the discoverability of natural laws. "To God," he wrote, "there are, in the whole material world, material laws, figures and relations of special excellency and of the most appropriate order." He affirmed that "those laws are within the grasp of the human mind," because "God wanted us to recognize them by creating us after his own image so that we could share in his own thoughts."[17] Kepler's scientific ethos had biblical foundations: humans are guided toward truth by factors that are grounded in God's attributes, God's creation, and God's image in us as investigators. Because humans were created in God's image, Kepler argued, we can grasp some of the same mathematical ideas that come from the mind of God.

Furthermore, Kepler employed Christian theology to uproot Aristotle's and Bruno's idea of an eternal cosmos animated by a world soul. He thus cleared the way for superior scientific viewpoints. Such a Christian worldview was evident in a 1605 letter Kepler wrote about his research that unveiled the first two laws of planetary motion:

> I am now much engaged in investigating physical causes. My aim is to say that the celestial machine is not like a divine animal but like a clock (and anyone who believes a clock has a soul gives the work the honor due to its maker) and that in it almost all the variety of motions is from one very simple magnetic force acting on bodies, as in the clock all motions are from a very simple weight.[18]

Kepler theorized that the sun exerts a force on the planets that keeps them orbiting in elliptical paths. He used a clock metaphor to distinguish his cosmology from Aristotle's idea of planets moving by means of a world soul. But Kepler did not envision a strictly mechanical view of a cosmos that could run without God's providence. He even speculated about a nonmechanical soul in the earth itself, which he thought helped explain weather patterns by means of mathematically traceable celestial events.[19] He was wrong about that. But he succeeded in discovering laws of planetary motion by inspiration from the idea that God's regular manner of managing the cosmos comes by means of mathematical rules that planets obey.

Kepler contrasted his integrated mathematical-physical analysis of the universe with the view of Aristotle, "who did not believe that the World had been created and thus could not recognize the power of these quantitative figures as archetypes [i.e., mathematical design plans for the material world], because without an architect there is no such power in them to make anything" physical. Yet, Kepler continued, a mathematical study of celestial physics "is acceptable to me and to all Christians, since our Faith holds that the World, which had no previous existence, was created by God in weight, measure, and number, that is in accordance with ideas coeternal with Him."[20] Kepler proposed that among the mathematical ideas that exist in the divine mind, God freely selected some of them to govern his creation. Because God has the freedom to make many possible universes consistent with his eternal attributes, one cannot simply deduce from prior principles a single way that God must have created the world. Consequently, detecting likely truth in scientific ideas requires testing, sometimes with experiments, so as to reveal the virtues and vices of theories. Notice how Christianity helped construct the foundation for scientific methodology.

The theologian-historian Christopher Kaiser, who also earned a PhD in astrogeophysics, argues that "two complementary features in Kepler's belief structure" guided his scientific discovery:

> The belief that the universe is ordered by mathematical laws is one of these.... Second is the belief that mortals have the intelligence needed to discover those laws because they are created in the divine image.... These two beliefs correspond to the dual

belief identified by modern physicists like Einstein and Paul Davies as the foundation of their scientific work. There is a correspondence between the depths of the human psyche and the deep structures of the universe.[21]

Kepler's basis for believing in a lawful and discoverable universe is worth investigating further. By digging deeper in earlier history, we can dispense with a common misconception.

The Deeper Judeo-Christian Foundation for Natural Laws

Stephen Hawking's *A Brief History of Time* (1988), the second bestselling science book of all time (after Desmond Morris's *The Naked Ape*), is riddled with common misconceptions about the history of science and religion. According to Hawking, the earliest explanations of the cosmos invoked unpredictable and capricious spiritual beings as the cause of natural phenomena. Although this is true of many cultures, the Judeo-Christian tradition never promoted this ideology, despite Hawking's suggestion otherwise. Hawking claimed, "Gradually, however, it must have been noticed that there were certain regularities." Sagan and Tyson similarly mischaracterized history in *Cosmos*.

Hawking, Sagan, and Tyson seem unaware that belief in the Judeo-Christian God actually *supported* the idea that the universe is predictable and knowable as a law-abiding system, which is foundational to science. In this case theology got it right first, and then successful scientific endeavor followed. Kepler's belief in cosmic lawfulness traces back to the Hebrew scriptures. The same is true for countless other scientists. Here is one of the key texts: "Thus says the LORD, who gives the sun for light by day and the fixed order of the moon and the stars for light by night.... If this fixed order departs from before me, declares the LORD, then shall the offspring of Israel cease from being a nation before me forever" (Jeremiah 31:35–36, ESV). Christopher Kaiser notes:

> The term translated here as "fixed order" (NRSV) is the Hebrew word, *hoq*, meaning a royal decree or law. It is translated as *nómos*, the Greek word for law, in the Septuagint, and as *lex* in

Jerome's Latin translation, the Vulgate. The biblical and theo-logical use of these terms played a huge role in the development of the idea of cosmic natural law inherited by modern science.[22]

Building on this Judaic foundation, Theophilus, bishop of the Syrian city of Antioch (died ca. 185), recognized that "an earthly king is believed to exist...by his laws." Although most people never see a king in person, they can infer his existence by observing the orderly society that such a king governs. Similarly, Theophilus argued, God is known "by his works," including "the timely rotation of the sea-sons...the various beauty of seeds, and plants, and fruits," and vari-ous "species of quadrupeds, and birds, and . . . the instinct implanted in these animals."[23] Drawing from their Jewish intellectual roots, early Christians recognized that the lawful natural world is evidence for a God who reigns supreme as the maker and ruler of all things.

Around the same time as Theophilus, Athenagoras of Athens argued that the repeating patterns of nature reflect God's *logos* (Greek, meaning "rationality" or "word"). He wrote that "the gen-eral constitution of nature" is ruled "by the law of reason" such that "there is nothing out of order" among natural things, and "each one of them has been produced by reason, and that, therefore, they do not transgress the order prescribed to them" by God.[24] This perspective echoed the opening of John's Gospel: "In the beginning was the *logos* and the *logos* was with God and the *logos* was God." John explains that this *logos* was Jesus of Nazareth, and that "all things were made through him."[25] He is the source of all good things, including a lawful universe that can be known by rational human minds. Medieval and early modern scientists commonly advocated this ancient theological foundation for scientific endeavor.

Kepler Pioneered the Integration of the History and Philosophy of Science

Kepler accomplished another amazing feat. In 1600, before discov-ering his three laws of planetary motion, he wrote a defense of the human ability to discover truth in astronomy. This came in response to skeptics who doubted that humans could ever know which system of astronomy is actually true: geocentric, geoheliocentric (Tychonic), or heliocentric. In so doing, Kepler pioneered the systematic study of the traits of a good theory, which are now called "theoretical virtues." He did this by integrating the history and philosophy of science. "As in every discipline," he wrote, "so in astronomy also, the things which we teach the reader by drawing conclusions we teach altogether seriously, not in jest. So we hold whatever there is in our conclusions to have been established as true."[26] Kepler thought that truth about the world is discoverable for two main reasons: God made us in his image and God designed the world to be discoverable by intelligent embodied creatures like us.

Nicholas Jardine's translation and interpretation of Kepler's remarkable achievement in the analysis of theory virtues is contained in a book boldly titled *The Birth of History and Philosophy of Science* (1984). Scholars in various fields had long before recognized that true accounts of the world would be marked by evidential accuracy, simplicity, and several other theory virtues. But Kepler was the first (as far as we know) to employ systematically the history and philosophy of science to characterize the virtues of a good theory.[27] Attention to such theoretical virtues has proved to be a very successful strategy for the progress of science.[28]

Aware of the roughly equivalent evidential accuracy of the Copernican system compared to its rivals (especially the Tychonic system), Kepler appealed to additional theoretical virtues as likely indicators of truth. "False hypotheses, which together yield the truth by chance, do not...retain the habit of yielding the truth, but betray themselves."[29] Here Kepler referred to the *failure* of a theory to instantiate the virtue of durability. Put positively, a theory exhibits durability if it maintains the habit of having its predictions confirmed. Note how Kepler recognized the history of science as essential to the practice of the philosophy of science. The theoretical virtue of durability is impossible to identify without doing your historical homework. The durability of a theory provides some confidence that our minds are probably tracking with the mind of God, who is the cause of the natural world and its laws. That's how Kepler thought of it.

Kepler's defense of scientific realism—that science progressively discovers truth—also called on the theoretical virtue of *causal adequacy*: "Even if the conclusions of two hypotheses coincide in the geometrical realm, each hypothesis will have its own peculiar corollary in the physical realm."[30] Here Kepler urged astronomers to identify causes that plausibly account for the physical details of planetary motion, rather than just "save the phenomena." To "save the phenomena" in astronomy usually meant merely to achieve evidential accuracy by means of geometrical hypotheses divorced from causal-physical considerations.[31] Kepler called his approach to astronomy "celestial physics" and managed to discover three laws of planetary motion by means of it. Attention to the causal adequacy of theories was one of the distinguishing marks of his celestial physics. So it has been ever since.

Kepler thought that beauty was another trait of a likely true theory, especially when it comes in the form of simplicity. This made sense to him because God "introduced nothing into Nature without thoroughly foreseeing not only its necessity but its beauty and power to delight."[32] He toyed with many possible beautiful, closed, curved figures to see which would best fit the astronomical data of planetary positions. He finally settled on the ellipse, which is elegantly simple (though not as simple as the traditional circular heavenly motions that Copernicus and Galileo had envisioned). It also achieved an unprecedented level of evidential accuracy and other theory virtues.

How Bruno Messed Up

Kepler's life and work helps further debunk a specific myth used to portray science and religion at war: the myth of Bruno as a martyr for science.

In 1610 Kepler published an open letter to Galileo about his amazing telescopic discoveries. This letter sheds light on how scientific discovery interacts with discussion of human importance and other matters relevant to Christianity, including issues related to Bruno's execution. Galileo had just discovered four of Jupiter's moons. Prior to this, in the Copernican system, only earth was thought to have a satellite moon. In the Ptolemaic system the moon was just one more planet going around the earth like all the others. So Jupiter's moons strengthened the case that earth and Jupiter should be classified together as fellow planets with multiple similar properties. Based on this slightly strengthened analogy, Kepler concluded that Jupiter was also inhabited by intelligent creatures:

> Well, then, someone may say, if there are globes in the heaven similar to our earth, do we vie with them over who occupies the better portion of the universe? For if their globes are nobler, we are not the noblest of rational creatures. Then how can all things be made for man's sake? How can we be the masters of God's handiwork? It is difficult to unravel this knot, because we have not yet acquired all the relevant information.[33]

Notice how Kepler introduced this string of questions with "someone may say" and ended with a wait-and-see attitude given our limited knowledge. This showed judicious restraint, though after this he launched what he called his own "philosophical arguments" for human exceptionalism in relation to cosmological discovery. Unlike Bruno, Kepler was careful to separate out such highly speculative philosophy from the more rigorous and quantitatively precise work that enabled him to discover three laws of planetary motion.[34]

Kepler criticized the much less restrained Bruno and other similar contemporaries who belonged to what Kepler called the Pythagorean sect of philosophers. He said that they "neither deduce their reasons from the senses nor accommodate the causes of things to experience but...immediately, and as if inspired (by some kind of enthusiasm), conceive and develop in the walls of their heads a certain opinion about the arrangement of the world."[35] By contrast, Kepler insisted that our ideas about nature must be accountable to evidential accuracy and other theory virtues.[36]

Kepler continued his critique of Bruno's Pythagorean philosophy: "This sect misuses the authority of Copernicus as well as that of astronomy in general."[37] How so? Bruno and his like-minded contemporaries extrapolated unreasonably from the newly enlarged Copernican cosmos to an infinite one. And "once they have embraced" the idea of an infinite universe, Kepler said, "they stick to it and they drag in by the hair [things] which occur and are experienced every day in order to accommodate them to their axioms."[38] In other words, they never rigorously tested their belief in an infinite universe by the evidence of sense experience. Kepler criticized this promiscuous approach to nature in his book *On the New Star* (1604), which was about the latest of a series of "new stars"—what are now recognized as supernovas (explosions of some stars). They looked like new bright stars that appeared rapidly and then faded slowly. Kepler explained what he thought this sect of philosophers was doing inappropriately in this case:

> It pleases these philosophers to want this new star and all others of its kind to descend little by little from the depths of nature, which, they assert, extend to an infinite altitude, until according to the rules of optics it becomes very large and attracts the eyes of men; then it goes back to an infinite altitude and every day [becomes] so much smaller as it moves higher.[39]

Kepler complained that philosophers like Bruno could force whatever data appeared into their theory while leaving their original theory essentially untested. This was not a recipe for confidence.

Not everything that Bruno believed about nature was untestable. In some cases his beliefs failed crucial tests. For example, in his letter to Galileo, Kepler criticized Bruno's thermodynamic theory of planetary motion (as did we in chapter 4), noting that it was guilty of another kind of "weakness of reasoning." He pointed out how Galileo's telescopic discovery that Jupiter had four moons orbiting it violated Bruno's alleged "law of nature" according to which cold bodies (nonsuns) could move only around hot bodies (suns).[40] No wonder that Kepler referred to Bruno as the captain of a "dreadful philosophy" of little value to science.[41] Kepler acknowledged that Bruno got a few isolated facts about nature right but reminded Galileo, "Nevertheless, let him not lead us on to his belief in infinite worlds."[42]

Kepler expressed disgust over Bruno's execution, but he recognized that this man was burned alive for his pantheistic infinite universe (and more), not for a scientifically testable idea. Indeed, Bruno's infinity of worlds fell woefully short of Kepler's rigorous standard of testability. Even so, Kepler made clear that he did not view belief in ET as heresy. He and other Christian believers thought the stars (other suns) were inhabited: "In my opinion there is also humidity on the stars ... and therefore living creatures who benefit from these conditions. Not only the unfortunate Bruno who was burned in Rome on red hot coals, but also the venerated Brahe was of the opinion that there are living creatures on the stars."[43] Actually, Tycho probably did not believe in ET, but the fact that Kepler thought he did, while the elder astronomer clearly remained within theological orthodoxy, was the point of importance here.[44] Underscoring his opinion that believing in many worlds was nonheretical, Kepler introduced Bruno's name in his open letter to Galileo alongside that of "Cardinal [Nicholas] of Cusa," both of whom, he reported, believed that each star hosted other planets like our own solar system.[45]

Kepler got the Bruno story right, and he expressed more of this correct assessment in a letter to his friend Johann Georg Brengger: "I heard from Mr. Wackher that Bruno has been burnt in Rome; he is said to have been unyielding during the execution. He maintained the futility of all forms of religion and transformed the divine being into the world, into circles and points."[46]

Kepler essentially argued that Bruno's God was too small. The all-knowing and maximally loving divine creator as revealed in Scripture far exceeds in glory the world and its mathematical traits. The mathematical rules embedded within the world are themselves traceable back to the beautiful mind of God. Bruno's scientific musings had little bearing on his troubles with Rome. Max Caspar, Kepler's leading biographer, observed that Bruno's speculations "did not grow in the soil of astronomical research but originated from theological speculations and a pantheistic interpretation of nature. The idea of an infinite space is not rooted in experience but in metaphysics."[47] Bruno's god was an allegedly infinite universe. Kepler considered this god dead on arrival.

A Debt to Kepler

Getting acquainted with Johannes Kepler goes a long way toward dispelling the popular opinion that science and Christianity are typically at war with each other. This pioneering astronomer—one who promoted the Copernican system—was a man of intense personal piety whose theology guided his monumental astronomical discoveries. In *Epitome of Copernican Astronomy* and in many other writings, Kepler placed the Christian worldview at the foundation of science.

His writings also help defeat the later revisionist story that Bruno's execution in Rome represented a case of scientific martyrdom.[48]

Especially in *Epitome*, Kepler established a paradigm for the harmony of Christianity and science. As we will see in the next chapter, the vast majority of English-language astronomy textbooks followed that paradigm—at least until the nineteenth century, when Christianity-bashing myths became more widespread.

11

REMEMBERING AMERICA'S HARMONY OF SCIENCE AND FAITH

The heavens declare the glory of God.... The sun...runs its course with joy. Its rising is from the end of the heavens, and its circuit to the end of them....
—*Psalm 19:1–6, ESV*

In the lowest room of the World, is placed the sun.
—*Zechariah Brigden, 1659* New England Almanack

E mulating Kepler and other prominent early modern scientists, American astronomy education was founded on an understanding of the harmony of science and Christianity. The story begins at Harvard, America's first college, founded in 1636.

Henry Dunster (1609–1659), as Harvard's first president in 1640, taught the Ptolemaic geocentric system of the world. For this purpose he used one of the earliest natural philosophy textbooks adopted in America: Johannes Magirus's (1560–1596) *Physiologia peripatetica* (1610).[1] Although sometime between 1638 and 1723 Harvard's library acquired Kepler's textbook *Epitome of Copernican Astronomy* (1618–21), there is no record of whether instructors required students to read it.[2] But in the mid-seventeenth century, a new generation of Copernican astronomy textbooks arose that bear the marks of Kepler's influence.

Copernican Astronomy Goes to Harvard

The English astronomer Vincent Wing (1619–1668) wrote *Astronomia instaurata*,[3] which served as one of the earliest Copernican books Harvard students owned and read.[4] Although its formal role in the Harvard classroom is uncertain, this astronomical reference work provided the basis for a pro-Copernican essay that appeared in the 1659 *New England Almanack*. Zechariah Brigden, a Harvard graduate and tutor, edited the 1659 almanac for Harvard's press.[5] He credited Wing's book for his argument supporting the Copernican system.[6] Brigden's Copernican essay in that almanac marked one of the earliest scientific treatises printed in the English colonies.[7] America's first printing press, which arrived in Cambridge in 1638, may have been used to produce Brigden's 1659 almanac.

After mentioning the three chief systems of the world (Ptolemaic, Tychonic, and Copernican), Brigden defended Copernican astronomy with words that defy the historical myth that earth's Copernican decentralization was originally perceived as a human demotion: "In the lowest room of the World, is placed the sun, which challengeth to it self a centrall motion, finish't in the space of about 26 dayes."[8] Besides noting the sun's axial spin (every twenty-six days, as then estimated), notice Brigden's description of the sun's central location. The cosmic center, rather than treated as an honorable place from which earth was demoted, was thought to be the "the lowest room of the World." He considered locations higher up from the center of the universe to be privileged.

Ironically, the Pulitzer Prize–winning historian Samuel Eliot Morison (1887–1976),[9] the scholar who in 1934 published the first complete reprint of this textbook-inspired colonial Copernican essay, advanced the Copernican demotion myth. Referring to Copernicus and other prominent early modern scientists, Morison proclaimed, "They robbed the Earth of her traditional place as center of the universe, man of his proud preëminence as the creature for whose satisfaction all else was created, and the Almighty of his wire-pulling function in a snug little cosmos."[10]

Kepler's astronomy textbook and Brigden's early American defense of Copernican astronomy both undercut Morison's claims about the alleged anti-theistic implications of the new astronomy.

Furthermore, Wing's *Astronomia instaurata* (1656), the Copernican work that Brigden popularized in the 1659 almanac, contained nothing that substantiated the Copernican demotion myth.[11] So Brigden's Copernican essay demonstrates that at least some mid-seventeenth-century Harvard students retained a premodern valuation of the cosmic center as the "lowest room of the World." The Copernican demotion myth began to appear in popular literature in Europe around this time. Apparently this cultural virus had not yet spread to Harvard.

Copernican Textbooks in the Harvard Library

The printed Harvard College library catalogues up through the 1790 edition listed many of the textbooks to which students had access during the first century and a half of Harvard's existence. I found nine English-language astronomy textbooks and reference works, dating from 1656 to 1769:[12]

1. Vincent Wing, *Astronomia instaurata*, 1656
2. Joseph Moxon, *A Tutor to Astronomy and Geography*, 1665
3. David Gregory, *The Elements of Astronomy*, 1715
4. William Whiston, *Astronomical Lectures*, 1715
5. John Keill, *An Introduction to the True Astronomy*, 1721
6. Isaac Watts, *The Knowledge of the Heavens and the Earth Made Easy*, 1726
7. Roger Long, *Astronomy*, 1742
8. James Ferguson, *Astronomy Explained Upon Sir Isaac Newton's Principles*, 1756
9. William Emerson, *A System of Astronomy*, 1769

These textbooks exhibit some striking features in their treatment of science and religion. They promote neither the Copernican demotion myth nor any of the other myths about warfare between science and Christianity that we have surveyed. They convey a harmonious relationship between science and theistic religion. So they all followed the paradigm established by the first Copernican astronomy textbook: Kepler's *Epitome*.

In *Tutor to Astronomy and Geography* (1665), the English printer and mathematical instrument maker Joseph Moxon (1627–1700) argued for the Bible's compatibility with heliocentric astronomy. Moxon's biblical exegesis closely resembles the widely read argument for the biblical acceptability of Copernicanism in Kepler's *Astronomia nova* (1609).[13] Moxon published his book shortly after the first English translations of several key texts by Galileo and Kepler appeared, outlining ways of reconciling the Bible with Copernican astronomy.[14]

Citing many of the same Psalms that Kepler addressed, Moxon challenged critics of Copernicanism. He noted that those critics, "to prove the Earth's stability," cited Psalm 104:5, which says, "He set the Earth upon her Foundations, so that it shall never move." He also noted that they cited Psalm 24:2, which says of the earth, "He hath founded it upon the Seas, and established it upon the Floods." But Moxon pointed out that "there is nothing in these Scriptures against the Motion of the Globe of Earth." Rather, these passages were intended to communicate "that the Element of the Earth, I mean the Land, shall never move out of the place God hath assigned it."[15] He further explained that neither the psalmist nor any other biblical writer intended to "deny, that as it [earth] is a Globe, it hath a Motion." Instead, the biblical authors intended to teach that as earth "is a solid Substance, and Land, it hath its Foundations." So "it is plain the Spirit of God in those Texts means the Land, and not the whole Ball Astronomers call Earth, which consists both of Earth, Water, etc." Moxon observed that Moses used the word *earth* to mean several things, depending on the context, including just the dry-land portion of earth's surface as in Genesis 1:10: "God called the dry Land Earth."[16] If Moses could use the word *earth* to refer to "dry land," then the psalmist could do this for similar reasons. All of this is compatible with earth as an orbiting planet.

Similarly, Kepler in his *Astronomia nova* noted that some questioned Copernicanism by arguing that "Psalm 104, in its entirety, is a physical discussion, since the whole of it is concerned with physical matters." Kepler replied: "Nothing could be farther from the psalmist's intention than speculation about physical causes. For the whole thing is an exultation upon the greatness of God, who made all these things." The psalmist "treats the world" as "it appears to the eyes," not attempting to teach the counterintuitive truth of a moving earth. "If you consider carefully, you will see that it is a commentary upon the six days of creation in Genesis," Kepler advised.[17] He, like Moxon, alluded to how Moses in Genesis 1:10 used the word *earth* to mean just the dry-land portion of earth's surface. It was "earth" in this narrow sense that Psalm 104 celebrated—a stable place for human habitation. "For it is still true that the land, the work of God the architect, has not toppled as our buildings usually do, consumed by age and rot."[18] Kepler concluded his exegesis of Psalm 104 by reminding us that the psalmist "tells nothing that is not generally acknowledged, because his purpose was to praise things that are known, not to seek out the unknown. It was his wish to invite them to consider the benefits accruing to them from each of these works of the six days."[19]

Moxon also considered geocentric arguments that pointed to Psalm 19:5–6, "where the Psalmist speaking of the sun, saith, 'Which cometh forth as a Bridegroom out of his Chamber, and rejoyceth as a mighty man to run his Race: His going out is from the end of the Heaven, and his Compass is unto the ends of the same.'" But, he replied, the Holy Spirit here intended not to teach us whether the sun or earth actually moves but rather "to set us admiring the Works of the Creation, that so we might be the apter to glorifie him." Finally, Moxon pointed out that "to our sensual appearance the sun doth Rise and Set, and should the Spirit of God have said the Earth turns about, it might so have confounded Vulgar Understandings, that the Praises, Glory and Honour due unto God, might have been neglected through the Unbelief of that Truth which seems so improbable to our Sense of Seeing."[20]

In *Astronomia nova*, Kepler made the same case, arguing that Psalm 19 accurately described human experience. The psalmist "should not be judged to have spoken falsely," he said, because "the perception of the eyes"—that the sun moves—"also has its truth, well suited to the psalmodist's more hidden aim, the adumbration of the Gospel and also of the Son of God."[21] On the same note, Kepler (like Galileo a few years later) recommended a general rule of biblical interpretation: When the Bible describes "common things" (natural phenomena) "concerning which it is not their purpose to instruct humanity," they "speak with humans in the human manner, in order to be understood by them. They make use of what is generally acknowledged, in order to weave in other things more lofty and divine."[22]

While some other seventeenth-century astronomy textbooks echoed Kepler's and Galileo's arguments for the compatibility of the Bible with Copernican astronomy, eighteenth-century textbooks largely dropped this practice. Astronomy textbooks in the wake of Newton's *Mathematical Principles of Natural Philosophy* (1687) could make a persuasive scientific case for the Copernican system. From that point on, heliocentric astronomy was well established. This, coupled with earlier widely circulated arguments for the compatibility of the Bible with Copernicanism, made it unnecessary to make the biblical case anymore.

The Scottish mathematician David Gregory's *Elements of Astronomy* appeared in 1715 as a posthumous English translation of the 1702 Latin edition. Gregory (1659–1708) was an early defender of Newtonian theory within the classroom. His textbook aimed at making more accessible the "Celestial Physics, which the most sagacious Kepler had got the scent of, but the Prince of Geometers Sir Isaac Newton, brought to such a pitch as surprises all the world."[23] Gregory acknowledged his dependence on Kepler many times throughout his two-volume textbook.[24] He promised to "explain the Keplerian Physics delivered in... *Epitome of the Copernican Astronomy*, and that for the most part in his own Words."[25] Gregory updated Kepler's work in the light of Newton's grand synthesis, and never disputed the theistic worldview that accompanied the scientific achievements of Kepler and Newton.

William Whiston (1667–1752), like Gregory, leveraged his friendship with Newton to advance his career as a university professor of Newtonian physics. But after he publicly promoted Newton's anti-Trinitarian views in 1708, Whiston lost the Lucasian professorship at the University of Cambridge that he had acquired after Newton's 1701 resignation. Soon thereafter he published his *Astronomical Lectures* (1715), which he had delivered at the University of Cambridge. He frequently mentioned Kepler, Newton, and Gregory. Although a theistic worldview was evident in Whiston's teaching, he did not articulate it as explicitly as Kepler did a century earlier.

Note, however, Whiston's comments on the size of the cosmos and our place within it. The immensity of the universe that the telescope revealed posed no threat to Whiston's theism. The "vast Number of the Stars can seem strange to none," Whiston wrote, if we consider "the inexhaustible Power of Almighty God."[26] Whiston also expressed a modern view of the sun's place of importance. He spoke of the "Prince himself of our System, the Body of the sun," and contrasted this with the sun's "Planetary Attendants," of which earth was one. But like Kepler, Whiston acknowledged the importance of earth as a platform for observation that was well suited for science.[27]

John Keill (1671–1721), another notable English astronomer, also expressed appreciation for earth as a platform for observing and measuring the cosmos. He argued that a motionless earth would provide only "one Point or Corner" from which we could observe the universe, whereas our moving earth allowed us to obtain "a distinct Knowledge of this immense Palace of *God Almighty*, and have an *Idea* or Image of it impressed on our Minds, which is worthy of its Infinitely wise *Architect*."[28] Keill also cited several Psalms to convey a theistic perspective of the cosmos, including Psalm 19: "The heavens declare the glory of God."

The hymn writer, theologian, and logician Isaac Watts (1674–1748) wrote the astronomy textbook *The Knowledge of the Heavens and the Earth Made Easy* (1726). He introduced astronomy with reference to Psalm 8:3: "If we look upward with David to the Worlds above us, *we consider the Heavens as the Work of the Finger of God, and the moon and the Stars, which he hath ordained.*" Watts then remarked: "Nor was there ever any thing that has contributed to enlarge my Apprehensions of the immense Power of God, the Magnificence of his Creation, and his own transcendent Grandeur, so much as that little Portion of Astronomy which I have been able to attain."[29] He cited, for example, "the Motions of the Planetary Worlds," which were "governed and adjusted by... unerring Rules."[30] He then employed Psalm 8 to provoke humble reflection on our physically small but spiritually important role in God's plan: "When we muse on these things we may lose ourselves in holy Wonder, and cry out with the Psalmist, *Lord what is Man that thou art mindful of him, and the Son of Man that thou shouldest visit him?*"[31]

University of Cambridge astronomer and theologian Roger Long (1680–1770) wrote *Astronomy in Five Books* (1742), which offered a similar perspective on humanity's place in the cosmos. He alluded to Psalm 8 as an antidote to the "pride of man," which inclined us to think that "all things were made for his use." Although God gave us "dominion" over earth, the rule of humanity "reaches no farther," he asserted, alluding to Psalm 8:6–8, where the extent of human dominion appears limited to earth. Long said that in observing the cosmos, the psalmist "is humbled into an admiration of the great condescension of the author of such wonderful works, that he should vouchsafe to cast an eye of regard upon man, and make him the object of his favour."[32] So we are small but significant.

Long concluded by calling astronomy "the most delightful, the most extensive, and the sublimest science which the great Author of Nature has held forth for employment to the faculties of man: the study of which, perhaps first led mankind to the true knowledge of his great and allwise creator."[33]

James Ferguson (1710–1776), famed self-taught popular lecturer and astronomical gadget maker,[34] wrote *Astronomy Explained upon Sir Isaac Newton's Principles* (1756), which became a widely used text in colonial America and the early republic.[35] In the book Ferguson addressed an issue that Newton himself never settled: gravity's relationship to God and whether gravity is "mechanical or not." He concluded that gravity "seems to surpass the power of mechanism" and so is "either the immediate agency of the Deity, or effected by a law originally established and imprest on all matter by him."[36] Additional theological matters crop up in Ferguson's textbook, including biblical chronology, the astronomical conditions that obtained on the day of Christ's crucifixion,[37] and a prediction of the existence of intelligent God-worshipping extraterrestrials in many other worlds.[38]

One can detect a shift in theological orientation in *A System of Astronomy* (1769) by William Emerson (1701–1782). In his preface, Emerson introduced astronomy in this manner: "By the wonderful discoveries made in this science, we can comprehend the vast extent of the universe, and the amazing beauty of its structure; which gives us a grand and magnificent notion of the world and all its parts; at the same time that we see the Earth which we inhabit, is but a poor, small, trifling part of the whole."[39] Whereas earlier astronomy textbook authors in our survey typically quoted the Bible to recognize human significance in a vast cosmos, Emerson did not appeal to scriptural authority and instead argued that astronomy pointed to "the wisdom and power of the Deity." He noted that "mere mechanical laws could never give rise" to all the motions of the universe. Those motions were "so nicely adjusted to one another," showing that "they were all originally contrived and executed by some powerful being, acting with council and design, which demonstrates the profound skill of the great architect."[40] Emerson expressed a deistic outlook rather than a traditional Christian one.

This foreshadowed a much more dramatic shift to come in science education in the centuries ahead.

Colonial Scientists Promoted Kepler's Harmony

At a time when anti-theists like Carl Sagan, Stephen Hawking, and Neil deGrasse Tyson have done so much to shape the popular understanding of the cosmos—and to advance the myths this book debunks—it is difficult even to conceive that science educators could envision a harmonious relationship between science and Christianity. But that is precisely what most astronomy textbooks did in early America.

Although eighteenth-century astronomy textbooks made fewer theological arguments than Kepler had in his technical monographs and in his textbook *Epitome*, the theology they promoted always harmonized with science. That harmony is difficult to find today. It is worth reclaiming.

12

TELLING THE
LARGER TRUTH

The greatest myth in the history of science and religion holds
that they have been in a state of constant conflict.
> —*Ronald L. Numbers,* Galileo Goes to Jail: And Other
> Myths About Science and Religion *(2009), 1*

Historical myths about science hinder science literacy and
advance a distorted portrayal of how science has been—and is—
done.
> —*Ronald L. Numbers and Kostas Kampourakis,* Newton's
> Apple and Other Myths About Science *(2015), 1*

Among the stories scientists tell, discerning readers will distinguish the credible from the unbelievable. The unbelievable narratives reveal much about the prejudices of certain scientists. This calls for investigation into the unique features and common patterns on display in the myths many scientists believe.

The early chapters of this book debunked six false stories that made Christianity look as if it has been incompatible with science throughout history. The "big myth" alleged that educated premodern people in the Western tradition thought that the universe was small—a cozy little place just for human benefit—and that when modern science showed otherwise, it dealt a blow to faith. This story gives the false impression that science progressively triumphs over religious faith. After the fake history is vanquished, philosophers still have work to do. Size and significance do not necessarily correlate. Sometimes tiny things are precious and big things are less valuable. Compare a human baby with 1 million cubic miles of interstellar gas and dust.

As chapter 11 demonstrated, the first two centuries of Anglo-American Copernican astronomy education overwhelmingly avoided the pseudo-history of the big myth, and all the other myths analyzed in my book. Furthermore, these astronomy educators often drew from the ancient wisdom of Psalms 8 and 19 to grasp the significance of tiny humanity on the cosmic scale. This trend tapered off during the nineteenth and twentieth centuries, but it has never been extinguished, even today.[1]

Here are two of the insights that astronomers over the past four centuries sometimes gleaned from those two ancient Hebrew songs. Cosmic immensity communicates God's glory, but his greatness exceeds the cosmos. Although humans are vanishingly small on the cosmic scale, God has made us very significant as his terrestrial representatives. The second verse of Psalm 8 takes on added significance in light of the history of science and theology in the Western tradition. Consider, in a cosmic context, the "infants" the psalmist refers to:

You have set your glory above the heavens.
Out of the mouth of babies and infants,
you have established strength because of your foes,
to still the enemy and the avenger.
When I look at your heavens, the work of your fingers,
the moon and the stars, which you have set in place,
what is man that you are mindful of him,
and the son of man that you care for him?
Yet you have made him a little lower than the heavenly beings
and crowned him with glory and honor.[2]

The idea is that God can take apparently insignificant persons (humanity, Israel, or the Christ child) and do great things through them that will silence his foes. Even if humanity on a cosmic scale is no more than an infant, overshadowed by perhaps older and wiser alien civilizations, that does not preclude human significance in a biblical sense. Many astronomers and theologians during the past four centuries have understood mankind's true significance in just this way.[3]

Storytelling Patterns in Science and Religion

Together with the big myth, the dark myth and the flat myth form a trilogy of distortions of *premodern* science and Christianity. The dark myth would have us believe that the medieval Catholic Church suppressed the growth of science, so that a shadow fell over the European intellect—the "Dark Ages." Carl Sagan's *Cosmos* TV series did a marvelous job of cultivating this erroneous view of Europeans living between about 500 and 1500. The flat myth, which is the most prominent and pernicious constituent of the dark myth, alleges that because of church-induced ignorance, European intellectuals before 1492 believed in a flat earth. These three stories that bash premodern Christianity are sure to make many anti-theists today feel quite good about themselves—until they hear the rest of the story. Over the past quarter century, I have particularly enjoyed watching my students squirm as they realized the sagacity of their medieval university counterparts. Compared to most college students today, medieval students could argue more extensively from observations to the conclusion that earth is round.

The second trilogy of historic myths has to do with *early modern* science and religion: Giordano Bruno as a martyr for science, the distorted Galileo affair, and the Copernican demotion of earth to cosmic insignificance. These myths largely originated in the mid-seventeenth century and were based on insufficient documentary evidence, exaggerations, anachronistic reconstructions of history, or some combination thereof. The table on the next page displays data from a sample of 130 astronomy textbooks distributed over four centuries. The myths are arranged in decreasing order of frequency of occurrence, showing that the myths about early modern history have been perpetuated more frequently than those dealing with premodern science and Christianity. None of the textbooks published before 1789 contained any of these myths.

Historic warfare myths in astronomy textbooks: 1618–2018		
Myth	Range of occurrence	Number of textbooks
Copernican demotion of earth	1849–present	26 (20 percent)
Distorted Galileo affair	1809–present	23 (18 percent)
Bruno as martyr for science	1897–present	15 (13 percent)
Dark Ages	1809–present	12 (9 percent)
Medieval flat earth belief	1818–present	8 (6 percent)
Premodern belief in a small cosmos	1789–1984	7 (5 percent)

If we eliminate pre-1789 textbooks (which include none of the myths) and eliminate textbooks having no content on the history of science (such textbooks have *no potential* for historic myth telling), then the above percentages come out to these values: 30, 26, 17, 14, 9, and 8. This also means that 60 percent of astronomy textbooks after 1789 that treat the history of science perpetuate at least one of the historic myths. Textbooks, though often corrupted, haven't fallen prey to these myths as easily as pop culture has—that is, until recently.

About 79 percent of currently used college astronomy textbooks contain at least one of these six myths. So textbooks now mirror popular culture in their very frequent perpetuation of the warfare myths. But even here there is *some* good news. As the table below demonstrates, textbooks published this decade traffic in some of these myths less frequently than earlier textbooks. In fact, the myth of the premodern belief in a small cosmos has been eliminated, while the Galileo myth has become less common. But these and other myths live on in popular culture.

The bad news is that, compared to earlier textbooks, current textbooks advance other myths more frequently. The Bruno myth has become more prominent, for example. And then there's the Copernican demotion myth. Earlier I noted that by the 1960s, this myth had become the dominant narrative in astronomy textbooks. That trend has escalated in textbooks currently in use.[4]

Historic warfare myths in astronomy textbooks: 2011–2018	
Copernican demotion of earth	10 (71 percent)
Bruno as martyr for science	5 (36 percent)
Distorted Galileo affair	2 (14 percent)
Dark Ages	2 (14 percent)
Medieval flat earth belief	1 (7 percent)
Premodern belief in a small cosmos	0

The Great Copernican Equivocation

How did the Copernican demotion myth become the dominant narrative in astronomy literature? Chaisson and McMillan's *Astronomy Today* (2018) offers a clue. After surveying overwhelming evidence for the Copernican system, the authors claim: "This removal of Earth from any position of great cosmological significance is generally known, even today, as the *Copernican principle*. It has become a cornerstone of modern astrophysics."

Unlike other science-religion conflict myths, the Copernican demotion has been canonized as an astronomical "principle" that bears the name of a scientific saint.[5] The rhetorical strategy is to equivocate between the now unassailable argument for a moving earth and the contested case for our cosmic insignificance. In the name of Copernicus, the evidence for both is declared "overwhelming." Most readers are no doubt unaware of this rhetorical sleight of hand.

Why the equivocation?

The cosmology literature expert Dennis Danielson suggests that in perpetuating the Copernican demotion myth, many scientists may be guilty of "more than just an innocent confusion." Rather, he suspects that the myth "functions as a self-congratulatory story" of "materialist modernism."[6] In any case, in evaluating the merits of the Copernican principle today, remember that Copernicus himself—indeed, most early modern astronomers—did not actually embrace the idea.

Trading Copernican Demotion for Alien Exaltation

One of the important findings of my book has been that astronomy textbooks and popular science literature have recently made a remarkable "trade-in" offer: Copernican demotion is being exchanged for an envisioned promotion for humanity through enlightenment delivered by extraterrestrial beings. The ET enlightenment story is an imaginative worldview-shaping narrative. Any ET capable of traveling to earth would be so advanced that their technology would be indistinguishable from magic, many scientists and philosophers now tell us. This encounter is envisioned to end human-centeredness and cosmic loneliness, leading to an age of universal spirituality beyond sectarian terrestrial religion. Christianity's "Incarnation," God taking on *human* flesh in Christ as the cosmic savior, is especially repugnant to such storytellers. But chapter 7 demonstrated that most of this ET enlightenment hype breaks down under scrutiny.

The ET enlightenment myth, the idea that alien knowledge would radically revise terrestrial worldviews, is worth tracking further in my sample of 130 astronomy textbooks. Of these modern textbooks spread over four centuries, 51 of them treat the existence of ET as plausible or probable. And of these 51, only 10 contain an additional ET enlightenment narrative. The mythic ET enlightenment story is mostly a recent addition to astronomy textbook lore.

The first instance I found of ET enlightenment in an astronomy textbook is worthy of special attention. Soon after retiring as director of the Hayden Planetarium in 1972, Franklyn Mansfield Branley wrote:

> There have been significant milestones in man's journey toward understanding who and what he is, the nature of his world, his place in the universe. Some examples are the heliocentric solar system of Copernicus.... But certainly the most mind-stretching discovery that man could ever make would be the existence of intelligent creatures at some location out among the stars. Upheavals of immense proportion would rock the foundations of many of men's religious beliefs. Tenets of schools of philosophy would have to be modified, and political systems reshuffled. This would be a changed world indeed.[7]

Meeting ET would be the greatest discovery of all time and would radically revise religion and philosophy. Branley did not reach this conclusion by scientific inquiry. It is merely an expression of his religious and philosophical opinions.

As a former planetarium director at a liberal arts university, I am familiar with the planetarium community to which Brantley belonged, as does a successor of his at the Hayden Planetarium, Neil deGrasse Tyson. This branch of science education is tilted toward storytelling. The aim is to educate and deeply move an audience under a starry dome through music, images, and narration. Planetarium shows often relate specific scientific content to the larger question: "What is our place in the universe?" Science is the right discipline for addressing matters of *spatial* location, but knowing *our place* calls for additional help from other fields, especially theology and philosophy. The same could be said of the two *Cosmos* TV series. Carl Sagan and Neil deGrasse Tyson have taught science, but also much more that is beyond their expertise and competence.

And thanks to recent astronomy textbooks, the ET enlightenment myth is reaching countless classrooms. In fact, the majority of textbooks in my contemporary sample—eight out of fourteen—contain this futuristic myth. These include some widely adopted titles, so collectively they command a high percentage of astronomy classrooms. To see the religious fervor with which textbook authors advance this myth, read the sidebar beginning on page 190.

Students should realize that astronomy professors will deploy the ET enlightenment story and other mythic elements of "the cosmic perspective" (as Tyson loves to call it) without identifying them as futuristic myth. It may seem that they are just teaching science. That makes the mythic components of such courses more difficult to detect and analyze critically. To begin correcting these delusions is the purpose of this book.

Materialism's New Magic

Materialism and magic are getting harder to distinguish, thanks to popular science writing about ET and AI. In a book that purports to defeat belief in God through materialistic philosophy and science, Richard Dawkins declares his belief in evolved aliens that are "superhuman, to the point of being god-like."[8] Similarly, "skeptic" Michael Shermer asserts that advanced ET would be "indistinguishable from God."[9] Technology wizard Ray Kurzweil predicts

Broadcasting ET Enlightenment Religiously

Imagine a TV show in which contemporary astronomers proclaim the good news of ET enlightenment. The *italicized* parts of this imagined broadcast are actual quotations from textbooks—ones currently used in teaching astronomy.[11]

As the lights go up, a half-bald astronomer with a ponytail, Dr. Benedict, strides to center stage: "We, and our alien siblings, are the voice of the cosmos. Discovering extraterrestrial life will *surely mark a turning point in the brief history of our species,* and *this turning point is likely to be reached within the next few decades or centuries.*"[12]

A colleague, Dr. Sedgwick, takes his turn: "Encountering ET *would end the self-centered isolation of humanity and stimulate a reevaluation of the meaning of human existence. We may never realize our full potential as humans until* the first contact with aliens."[13]

Professor Chaffinch interjects, "Yes, the day of ET will *have a profound effect on human culture.*"[14]

Professor Franklin chimes in: "*Our personal questions about life…, love… and meaning… are all… framed against the background of our place in the Universe.* So reevaluate with greater scrutiny the *other stories of the cosmos* that *have come before the one you will encounter* today. These other stories are *the great creation myths of cultures that preceded ours.*"[15]

Professor Chaffinch replies, "Thank you, brother Franklin, self-identified *evangelist of science,*[16] for your call to ponder the great stories of astronomy."

Franklin addresses the studio audience: "*How would people understand their religion if they suddenly found out that*

that soon magic resembling the world of Harry Potter will break forth as the AI-Singularity dawns.[10] But if aliens have already had their AI-Singularity, then godlike magic is already loose in the cosmos. Although chapter 7 debunked this AI-ET myth, an increasing

other intelligent beings existed who were not mentioned in any scripture? And if we were to fail to make contact with aliens indefinitely, would humanity be destined to spend its entire history knowing only its own perspective of life and death, meaning and existence?"[17]

Dr. Slaton asserts: "Receiving a message from an alien civilization could dramatically change the course of our own civilization, through the sharing of scientific information with another species or an awakening of social or humanistic enlightenment."[18]

Dr. Pasternak insists: "It would be one of the most important and mind-blowing discoveries ever, if not the greatest discovery of all time."[19] He continues: "But meanwhile, short of actual contact, we can console ourselves with wisdom from that late great celestial high priest, Carl Sagan, who essentially said that the Universe has found a way to know itself, through us. We are the observers and explorers of the Universe; we are its brains and its conscience. This makes each one of us special to the Universe as a whole."[20]

Dr. Faberge concludes the program with this testimony: "The study of astronomy takes in history as well as issues that border on the ethical and religious. Astronomy is like that: It offers a modern-day version of Genesis—and of the Apocalypse, too.... Take time out to contemplate the broader implications of what we are discussing tonight. This is one of the rare opportunities in life to think about who you are and where you and the human race are going."[21]

With that, the astronomical revival meeting ends. But the message will carry through in the countless astronomy classrooms where the astronomers' sermonizing texts are used.

number of people believe it on the authority of prominent anti-theistic scientists and technology leaders.

The growing belief in such materialistic magic helps make sense of a resurgence of the occult among science-educated millennials who say their religion is "none."[22] According to Pew's massive 2014 Religious Landscape Study, 35 percent of Americans age eighteen to twenty-nine are "nones"—those claiming no particular faith.[23] With diminishing belief in theism, and with growing enthusiasm for the AI-ET enlightenment myth, millennials are increasingly likely to justify their experimental spirituality on naturalistic or materialistic grounds. Many high-profile scientists and technologists since 1973 have affirmed Arthur C. Clarke's dictum about future technology's being "indistinguishable from magic."

Katie Baker, reporting for *Newsweek*, says that astrology and Tarot-card reading are the leading occult activities of American millennials. Astrologer Susan Miller told Baker that her horoscope website attracts six million unique visitors monthly, "73 percent of whom went to college or graduate school." Some millennials told Baker that the occult was a wonderful nonreligious way to connect with "the world" as "both older and bigger than they are."[24] This echoes Sagan's opening *Cosmos* monologue: "The size and age of the cosmos are beyond ordinary human understanding," and "Our contemplations of the cosmos stir us.... We know we are approaching the grandest of mysteries."[25]

Sociologists Joseph Baker and Christopher Bader have found a correlation between the proliferation of the "nones" and "the cultural diffusion of the paranormal."[26] Rather than embracing old-fashioned atheism, most nones aspire to be spiritual but not religious. Tyson appeals to this audience when he announces that "the cosmic perspective is spiritual—even redemptive—but not religious." Because the paranormal refers to phenomena currently unexplained by science, one might see this as the antithesis of science. But the AI-ET enlightenment narrative has partially closed the perceived gap between science and the paranormal.[27] Recognizing this story as a myth will go a long way toward decoupling it from the prestige of science.

Materialism's magic had lodged deeply in the heart of Carl Sagan. Like H. G. Wells, Sagan ended up finding religion in the attempt to escape it. Here is how one of Sagan's leading biographers put it: "Anyone who studies Sagan's life long enough recognizes its seeming contradictions:...its ruthless skepticism crossed with a love of mythology, dreams, and theology camouflaged as science fiction."[28] The futuristic mythology of sci-fi, especially Clarke's dictum about forthcoming technological magic, guided Sagan's imagination about humanity's destiny. Sagan replaced the Abrahamic God of his ancestors with anticipated materialistic magic as we join other enlightened galactic civilizations.

Ironically, many of the materialists and naturalists who criticize Christianity for hindering science with its supernatural claims have lately reintroduced the approximate equivalent of pagan magic by extrapolating technological evolution to inconceivable heights. Dawkins thinks alien civilizations are so advanced that they will "exceed anything a theologian could possibly imagine" about God.[29] No wonder science-educated adults are dabbling in the occult in growing numbers. ET and demonic powers are getting harder to distinguish in latter-day naturalism.

Will ET enlightenment be the final great myth about science and religion? Will the myth of alien techno-magic be irresistible for many people, despite the arguments against it in chapter 7?

The True Story After Materialistic Magic and the Historic Myths Are Removed

This book has debunked seven of the most pervasive and consequential myths about the history and future of science and religion. These myths all have something in common: they have been used to oppose the Christian faith more than any other religious faith. And yet all seven myths, in light of a perspective informed by sober science and accurate history, are unbelievable.

It is largely because of these seven myths that many people have failed to recognize Christianity's extensive positive contributions to science. The myths would have us believe that science came of age (and will continue to flourish) only in the victory of naturalism over Christianity. Yet as we have seen, Christianity has played an important role in the growth of science.

Christianity played a significant part in the development of experimental methods aimed at a closer reading of God's book of nature.

The Christian belief in divine freedom undercut the view, established by Plato and Aristotle, that the structure of the cosmos is a necessary one. Christians insisted that God could have created a universe quite different from the one Aristotle imagined, and so testing multiple hypotheses by experiment was an effective way to determine which set of natural laws God actually created to govern our cosmos.[30]

As we dig deeper into the foundations of science, we see that Christianity cultivated both humility and confidence in human knowledge. That confidence derived from the orderliness of God's world, designed for discovery by his human image bearers. Belief in God as the universal law giver encouraged investigation of nature to discover natural laws, as the remarkable achievements of Kepler and Galileo demonstrate. For this reason, Christianity could prove to be much *more* hospitable to the growth of science than are naturalistic or materialistic worldviews. The materialists' criticism of miracles in the Judeo-Christian tradition miss the mark. Those criticisms fail to recognize that the very notion of a miracle—a *rare* divine sign—would be inconceivable without the companion idea of nature's regularity.

At the same time, human fallibility was one of the most persistent themes in the Bible. The Christian doctrine of the Fall of Adam and Eve (and our status as finite creatures) provided an explanation for the difficulty of human reason in achieving certainty about the cosmos, with a consequent emphasis on the testing of hypotheses. Many medieval and early modern scientists embraced this balance of confidence and humility.[31]

Science and biblical religion have been friends for a long time. Friends sometimes fight, but in the case of strong relationships, conflict usually can be traced back to misinterpretations (including myths advanced by third parties with an interest in promoting disharmony). Judeo-Christian theology helped foster the ideas of experimental inquiry, universal natural laws, mathematical physics, and investigative confidence balanced with humility. Christian institutions, especially since the rise of the medieval university, provided a supportive environment for scientific inquiry and instruction.

Christianity, contrary to the mantra-like assertions from media and textbooks, is not "anti-science." Most of the founders of modern science were Christians, and that is no mere cultural happenstance. Belief in the biblical God provided crucial support to the notion that the universe is predictable and knowable, an idea that is foundational to science. Christianity also encouraged the balance of humility and confidence in knowledge characteristic of successful scientists such as Kepler.

A close study of the history of science shows that, far from inevitably standing in conflict with science, theistic religion nurtured the development of modern science from its start. If we were to rerun history and remove the influence of Christianity from the rise of modern science, today we would be living in a darker, more confused, and more ignorant world.

Appendix A

THE FIRST URBAN MYTH OF THE SPACE AGE

There is another kind of myth that we have not yet discussed: the urban myth. Consider the first major urban myth of the space age. The cosmonaut Yuri Gagarin (1934–1968) became the first man in space in 1961 and allegedly announced that he "did not see God" up there.

The story delighted many atheists on earth. But because there are inconsistent accounts of what he said (and when), and no available documentation to settle the matter, this anecdote is best classified as an urban myth—a recent, sensational, undocumented claim proliferated through popular communication channels.

This story first circulated almost immediately after Gagarin completed his historic flight. But decades later, in a 2006 interview, Gagarin's colleague Colonel Valentin Petrov said that the Communist Party fabricated the story. The Soviet Union, an officially atheistic state, saw a propaganda opportunity. According to Petrov, Soviet leader Nikita Khrushchev had originally claimed that Gagarin "saw no God" in space. Petrov, who in the 1960s was Gagarin's friend and fellow pilot, served as lecturer at the Gagarin Air Force Academy at the time of the interview.[2]

One of the most reputable Gagarin biographies, written by Andrew Jenks, does not address the Gagarin urban myth. Jenks does, however, show how Soviet officials carefully managed Gagarin and his public image—perpetrating numerous lies for political ends.[3]

The Gagarin myth remains absent from astronomy textbooks. Scientists, regardless of their religious viewpoint or nationality, generally would not regard the scientifically undisciplined comments of an orbiting Russian military pilot as consequential for astronomy. Nor would the story, even if true, establish anything of significance for theology. Scholars generally do not think that this sort of search for God would turn up anything relevant to the question of whether God exists.

But popular publications about science and religion often perpetuate this urban myth. Its popularity results, in part, from writers' frequent conflation of science with technology (as a pilot, Gagarin was a pioneer in the latter, not the former). New Atheist Victor Stenger contributed to such science-technology confusion in one of his most quoted post-9/11 statements: "Science flies us to the moon. Religion flies us into buildings."[4] Although the application of scientific theories constitutes one aspect of technology, much of technology (including most of aeronautics) has actually involved the empirical discovery of "know-how" knowledge without relying on particular scientific theories.[5]

The Gagarin urban myth did not replicate often within more disciplined literature, in part because able academics such as C. S. Lewis quickly and effectively refuted its significance. Noting Soviet reports that Gagarin had not found God in space, Lewis observed: "Looking for God—or heaven—by exploring space is like reading or seeing all Shakespeare's plays in the hope that you will find Shakespeare as one of the characters or Stratford as one of the places. Shakespeare is in one sense present at every moment in every play. But he is never present in the same way as a Falstaff or Lady Macbeth." Lewis then explained his analogy:

> My point is that, if God does exist, He is related to the universe more as an author is related to a play than as one object in the universe is related to another.... If God created the universe, He created space-time, which is to the universe as the metre is to a poem or the key is to music. To look for Him as one item within the framework which He Himself invented is nonsensical.... If God—such a God as any adult religion believes in—exists, mere movement in space will never bring you any nearer to Him

or any farther from Him.... Space travel really has nothing to do with the matter.... Those who do not find Him on earth are unlikely to find Him in space. (Hang it all, we're in space already; every year we go a huge circular tour in space.)[6]

Although the Gagarin tale may have originated from the top tier of the Communist Party as propaganda, it never gained much currency beyond the low culture of Main Street and Karl Marx Plaza. Regardless of whether it traces back to Khrushchev and the Central Committee, this popular story has functioned as an urban myth.

The warfare myths about science and religion that filter into astronomy textbooks and popular scientific literature appear more plausible to educated people than does the typical urban myth, and they are far more consequential. Urban myths are mostly just fun to tell and debunk. Beyond that, they don't matter much.

Appendix B

A NEW ARGUMENT FOR PASCAL AS A COPERNICAN

Chapter 1 introduced the renowned French mathematician, scientist, and philosopher Blaise Pascal. In his unfinished posthumous defense of Christianity, *Pensées*, he invited us to recognize earth "as a single speck compared with" its orbit around the sun. Or at least that is my interpretation of the passage. Most scholars have translated and interpreted this passage differently, taking it to indicate that Pascal was expressing a geocentric rather than a heliocentric view.

Why do I cite an infrequently used translation of the *Pensées* and treat Pascal as a Copernican when most have deemed him as uncommitted on these questions of cosmology? Because the most used English translation of *Pensées*, by W. F. Trotter, mistranslates part of the above passage. In Trotter's version, Pascal wrote, "let the earth appear to him a point in comparison with the vast circle described by the sun," rendering the last two words as "the sun" instead of "this star." The translation I have used, by Gertrude Burford Rawlings, is more accurate: "let the earth appear to him as a single speck compared with the vast orbit which *this star* describes."[1] My emphasis on "this star" is key, because grammatically it could point back to either the earth (indicating the Copernican system with a moving earth) or the sun (conveying the Ptolemaic idea of the sun moving around a motionless earth). The earlier part of this sentence, in Trotter's translation, describes the sun as "set like an eternal lamp to illumine the universe."[2] Trotter's faulty translation there misconstrues what "*this star"* refers to—Trotter selects *sun* (Ptolemaic interpretation), but Pascal almost certainly meant *earth* (Copernican interpretation). Copernicus in 1543 had reconceived earth so that it too became a "star"—a *planet* or "wandering star" (Greek πλάνης ἀστήρ, or *plánēs astēr*).

Trotter translates the final part of the passage from Pascal: "and let him wonder at the fact that *this vast circle* is itself but a very fine point in comparison with that described by the stars in their revolution round the firmament" (emphasis mine). "This vast circle" refers to whatever the earlier "vast circle" designates, which is the motion of either the earth or the sun. We can eliminate the latter possibility because, in the Ptolemaic system, the sun's circular path is *not* "a very fine point" compared to the size of the daily rotating sphere of fixed stars. According to the Copernican system, however, earth's circular path is indeed "a very fine point" compared to the sphere of fixed stars. As explained in chapter 1, Copernicus had to postulate this imperceptibly small earthly orbit in order to save his theory from the embarrassing lack of observed stellar parallax.

My Copernican interpretation of Pascal is further supported by O. W. Wight's 1859 edition of Pascal's *Pensées*. Wight, similar to Rawling's "this star" translation, renders it "that star," leaving open whether it refers to the earth or the sun as moving in a "vast circuit." But in a footnote, Wight reveals that in this part of the manuscript Pascal had crossed out the more clearly Copernican phrase "let the vast circuit which it describes make him regard the earth as a point" and replaced it with the French that underlies the above translations. Wight completes his footnote with this reaction: "In effacing this phrase, says M. Faugere, did Pascal wish to express no opinion on the systems of Copernicus and Galileo? This is certain, and 'that star' refers to the sun and not to the earth."[3] Wrong! What Pascal most likely did was initially express this passage in a more clearly Copernican manner but then decided (by striking the draft phrase above) to cloak his Copernican viewpoint in a way that only more careful and informed readers would grasp—a tactical move in the wake of Galileo's infamous trial (see chapter 5). Unfortunately, this clever tactic has fooled even scholars about the meaning of the passage.

The historian of astronomy Michael Crowe remained undecided on Pascal's attitude toward the Copernican system until 2013, when he wrote, "Pascal was indeed a Copernican." Crowe gave three reasons: "because a central theme of the *Pensées* is the infinitization of the universe, because no author before 1660 had written more effectively about the vastness of the universe, and because this vastness is a feature that results directly from the heliocentric and not from the geocentric system." Crowe concludes, "I suspect that Pascal writing on religious matters in his *Pensées* and adopting positions that upset many of his orthodox contemporaries did not wish directly to assert his Copernican convictions, but his genius and the geometry of his universe are definitely Copernican."[4]

I discovered this independently of Crowe and then found that Crowe had come to the same conclusion (though he does not cite the critical text above or explain its mistranslation and misinterpretation).

Notes

Introduction: The Truth Is Out There

1 Michael Witzel, *The Origins of the World's Mythologies* (Oxford: Oxford University Press, 2012), 7.

Chapter 1: Bigger Is Better

1 Bill Nye, https://www.youtube.com/watch?v=S4dZWbFs8T0.
2 Interview of George Abell by Spencer Weart, New York City, November 14, 1977, http://www.aip.org/history/ohilist/4475.html.
3 George Abell, *Exploration of the Universe*, 2nd ed. (New York: Holt-Rinehart and Winston, 1969), 34.
4 C. S. Lewis, *The Discarded Image: An Introduction to Medieval and Renaissance Literature* (Cambridge: Cambridge University Press, 1964), 97.
5 Ibid., 98–99.
6 Ibid.
7 Kepler to J. G. Herwart von Hohenburg, December 16, 1598, as translated in Carola Baumgardt, *Johannes Kepler: Life and Letters* (New York: Philosophical Library, 1951), 48–49.
8 Blaise Pascal, *Pascal's Pensées; or, Thoughts on Religion*, trans. Gertrude Burford Rawlings (Mount Vernon, NY: Peter Pauper Press, 1946), 104, frag. 72 (Brunschvicg system).
9 Ibid.
10 Pascal, *Pensées*, trans. W. F. Trotter (London: J. M. Dent; E. P. Dutton, 1908), frag. 206 (Brunschvicg system).
11 This is a distillation of Pascal based on my read of *Pensées*, trans. Trotter, cited by fragment number (Brunschvicg system) as printed in *Great Books of the Western World*, vol. 33 (Chicago: Encyclopaedia Britannica, 1952). Pascal explains that tiny sinful humans seek just about anything in the cosmos to fill the void inside us, including "stars, the heavens, earth, the elements, plants,...animals,...vices, adultery, incest." But "the infinite abyss can only be filled by an infinite and immutable object, that is to say, only by God himself" (frag. 425). Furthermore, "instead of complaining that God had hidden himself,"

205

he urges us to "give him thanks for not having revealed so much of himself" (frag. 288), which likely includes an allusion to the humble incarnation of Jesus conveyed in other fragments. After highlighting the opposite errors of pride (I am "by nature like" God) and total human debasement to animalistic lusts (my nature is "like that of the brutes"), Pascal points us to the solution to "who you are. Adam, Jesus Christ. If you are united to God, it is by grace, not by nature. If you are humbled, it is by penitence, not by nature. Thus this double capacity. You are not in the state of your creation.... Incredible that God should unite himself to us" (frag. 430) by the incarnation and redemption.

12   Christiaan Huygens, *The Celestial Worlds Discover'd* (translation of *Cosmotheoros*), 2nd ed. (London: James Knapton, 1722), 151.

13   Ibid., 10–11.

14   In the 1872 edition of his sermon "What Is Man?" (http://www.umcmission.org/Find-Resources/John-Wesley-Sermons/Sermon-103-What-is-Man), John Wesley challenged the case for ET and claimed that Huygens radically doubted ET shortly before his death in what became his posthumous *Cosmotheoros*. I thank Edward B. Davis for alerting me to Wesley. But Wesley cites only Huygen's case against lunar ET (and its broad implications) and does not evaluate Huygens's larger positive case for ET that is mostly independent of lunar considerations. Truth be told, Huygens's larger case for ET is mostly based on weak analogical arguments. For a fuller treatment of Huygens on this topic, see Michael J. Crowe, *The Extraterrestrial Life Debate, 1750–1900* (New York: Dover, 1999), 20–22.

15   William Whiston, *Astronomical Lectures, Read in the Publick Schools at Cambridge* (London: R. Senex, 1715), 38.

16   George Adams, *Astronomical and Geographical Essays* (London: William Young, 1789), 86–87.

17   "Adams, George (1750–1795)," in *Dictionary of National Biography*, as reproduced at https://en.wikisource.org/wiki/Adams,_George_(1750–1795)_(DNB00).

18   George G. Carey, *Astronomy as It Is Known at the Present Day* (London: William Cole, 1824), 26.

19   Newcomb published 318 items on astronomical subjects. See Marc Rothenberg, "The Educational and Intellectual Background of American Astronomers, 1825–1875" (Dissertation, University of Michigan, 1975), 100; William E. Carter and Merri Sue Carter, "Simon Newcomb, America's First Great Astronomer," *Physics Today* 62, no. 2 (2009).

20   Simon Newcomb, *Astronomy for Everybody* (New York: Macmillan, 1902), 9.

21   Newcomb wrote affirmingly about his Christian heritage: "My mother was the most profoundly and sincerely religious woman with whom I was ever intimately acquainted, and my father always entertained and expressed the highest admiration for her mental gifts." Simon Newcomb, *The Reminiscences of an Astronomer* (London: Harper, 1903), 6. He also remembered that upon reading Tobias Smollett's novel *The Adventures of Roderick Random* (1748) as a boy, Newcomb was overcome by "a feeling of horror that a man fighting a duel and finding himself, as he supposed, mortally wounded by his opponent, should occupy his mind with avenging his own death instead of making his peace with Heaven" (17). Finally, Newcomb had enduring respect for his Christian mentor Joseph Henry (1797–1878), the first secretary of the Smithsonian: "My talks with Professor Henry used to cover a wide field in scientific philosophy. Adherence to the Presbyterian church did not prevent his being as uncompromising an upholder of modern scientific views of the universe as I ever knew" (408). But in an anonymous essay "The Religion of To-day," *North American Review* 129 (Dec. 1879): 568, Newcomb wrote approvingly of a hypothetical confession of a new faith: "I have no belief in a personal Deity, ... in Christ as more than a philosopher, or in a future state of rewards and punishments. But I was born with a sense of duty to my fellow man." Newcomb's annotated copy of this article (indicating his authorship) is located in the Simon Newcomb Papers, Box 89, Library of Congress, according to Albert E. Moyer, *A Scientist's Voice in American Culture: Simon Newcomb and the Rhetoric of Scientific Method* (Berkeley: University of California Press, 1992), 264.

22 Richard Baum, "Flammarion, Nicolas Camille," in *The Biographical Encyclopedia of Astronomers*, ed. Thomas A. Hockey et al. (New York: Springer, 2014).

23 Camille Flammarion, *Astronomy for Amateurs*, trans., Frances A. Welby (New York: D. Appleton, 1904), 13–14. This was a translation of Flammarion's book for women: *Astronomie des Dames* (Paris: Ernest Flammarion, 1903). His earlier very popular book lacks the big myth: Camille Flammarion, *Popular Astronomy: A General Description of the Heavens*, trans. J. Ellard Gore (New York: D. Appleton, 1890). This was a translation of Flammarion's *Astronomie Populaire* (1879).

24 Flammarion, *Astronomy for Amateurs*, 16.

25 Ibid., 340.

26 G. K. Chesterton, *Orthodoxy* (London: John Lane the Bodley Head Ltd., 1900), 40.

27 The effectiveness of Lewis's refutation of the big myth was rooted in his expertise in medieval and early modern literature. This literary knowledge helped him to recognize how the contemporary philosophical error was entangled with the historical error of the "premodern small cosmos" story.

28 C. S. Lewis, "Dogma and the Universe," in *God in the Dock: Essays on Theology and Ethics*, ed. Walter Hooper (Grand Rapids: Eerdmans, 1970), 39. Originally published in *The Guardian*, March 19, 1943.

29 Ibid., 39.

30 Ibid., 39–40.

31 Ibid., 42.

32 Guy Kahane, "Our Cosmic Insignificance," *Noûs* 48, no. 4 (2014): 747–49.

33 Ibid., 767. I am ignoring the hermeneutically irresponsible anti-biblical discourse in Kahane's lengthy footnotes.

34 See http://hauskeller.blogspot.com/2016/11/guy-kahane-on-our-cosmic-significance.html. Similarly, British philosopher Nick Hughes argues: "I think that, contra Kahane, it is a sense of causal, rather than value, insignificance that is central to the sense that we are cosmically insignificant. Recognition of the tiny place we occupy in the Universe throws a stark light on our distinct lack of causal power." See Nick Hughes, "Do We Matter in the Cosmos?" *Aeon Magazine*, June 29, 2017, https://aeon.co/essays/just-a-recent-blip-in-the-cosmos-are-humans-insignificant.

Chapter 2: Idiots in the Dark

1 Carl Sagan, *Cosmos* (New York: Random House, 1980), 335.

2 Ibid., 332.

3 Ibid., 335.

4 David C. Lindberg and Michael H. Shank, eds., *Cambridge History of Science: Volume 2, Medieval Science* (Cambridge: Cambridge University Press, 2013), 10.

5 James Hannam, *God's Philosophers: How the Medieval World Laid the Foundations of Modern Science* (London: Icon Books, 2009).

6 Michael H. Shank, "Myth 1: That There Was No Scientific Activity between Greek Antiquity and the Scientific Revolution," in *Newton's Apple and Other Myths About Science*, ed. Ronald L. Numbers and Kostas Kampourakis (Cambridge, MA: Harvard University Press, 2015), 11.

7 Rodney Stark, *How the West Won: The Neglected Story of the Triumph of Modernity* (Wilmington, DE: ISI Books, 2014). Although Stark, a sociologist doing history, has read much recent work by historians of science, he sometimes extends his theses about the history of science beyond what the evidence supports. Nevertheless, his work deserves attention and refinement. Labeling him as a Christian apologist who is not a legitimate scholar is neither helpful nor true.

8 Lindberg and Shank, *Cambridge History of Science: Volume 2*, 2.

9   Jacob Burckhardt, et al., *The Civilization of the Renaissance in Italy* (London: Phaidon Press, 1944), Part II, ch. 1, 81. This book has appeared in countless editions since 1860.

10   John Ewing, *A Plain Elementary and Practical System of Natural Experimental Philosophy: Including Astronomy and Chronology* (Philadelphia: Hopkins and Earle, 1809), ix.

11   Ibid., xii

12   Ibid., xxii.

13   Ibid.

14   Ibid., 1.

15   Ibid., 2.

16   John Lauris Blake, *Conversations on the Evidences of Christianity* (Boston: Carter, Hendee, and Co., 1832), 36–45.

17   John Lauris Blake, *First Book in Astronomy Adapted to the Use of Common Schools* (Boston: Lincoln and Edmands, 1831), 10–11.

18   John William Draper, *History of the Conflict Between Religion and Science* (New York: D. Appleton, 1874), xi, 214–25.

19   I found only two science-religion warfare stories in Draper's science textbooks. In the astronomy section of his *Natural Philosophy* (physical science) textbook, he falsely claims that knowledge of a spherical earth "fell gradually into disrepute during the middle ages; it was reestablished at the restoration of learning only after a severe struggle." John William Draper, *A Text-Book on Natural Philosophy: For the Use of Schools and Colleges* (New York: Harper and Bros., 1847), 316. In the preface to his later *Human Physiology* textbook, he promised to treat his subject "as a branch of Physical Science," which meant eliminating "purely speculative doctrines and ideas, the relics of a philosophy (if such it can be called) which flourished in the Middle Ages, though now fast dying out." John William Draper, *Human Physiology, Statistical and Dynamical; or, the Conditions and Course of the Life of Man*, 2nd ed. (New York: Harper an Brothers, 1858), v.

20   George Abell, *Exploration of the Universe*, 2nd ed. (New York: Holt, Rinehart and Winston, 1969), 34. The first edition appeared in 1964.

21   Donald H. Menzel et al., *Survey of the Universe* (Englewood Cliffs, NJ: Prentice-Hall, 1970), 18–20.

22   Nicholas A. Pananides and Thomas Arny, *Introductory Astronomy*, 2nd ed. (Reading, MA: Addison-Wesley, 1979), 16.

23   http://www.adamfrankscience.com.

24   Adam Frank et al., *Astronomy: At Play in the Cosmos* (New York: W. W. Norton, 2016), 52.

25   Ibid., 44, 46.

26   David C. Lindberg, "The Medieval Church Encounters the Classical Tradition: Saint Augustine, Roger Bacon, and the Handmaiden Metaphor," in *When Science and Christianity Meet*, ed. David C. Lindberg and Ronald L. Numbers (Chicago: University of Chicago Press, 2003), 17; Bruce S. Eastwood, "Early-Medieval Cosmology, Astronomy, and Mathematics," in *Cambridge History of Science: Volume 2*, 305.

27   St. Augustine, *Contra Faustum Manichaeum* 32.20, as cited in Peter Harrison, "The Bible and the Emergence of Modern Science," *Science and Christian Belief* 18, no. 2 (2006): 118.

28   Eastwood, "Early-Medieval Cosmology, Astronomy, and Mathematics," 305.

29   Boethius and P. G. Walsh, *The Consolation of Philosophy* (Oxford: Oxford University Press, 2008), 34.

30   Ibid., 13 (the poem that begins chapter 5).

31   Ibid., 16 (the poem that begins chapter 6).

32   Eastwood, "Early-Medieval Cosmology, Astronomy, and Mathematics," 307.

33   Charles Burnett, "The Twelfth-Century Renaissance," in *Cambridge History of Science: Volume 2*, 379–81.

34   Walter Roy Laird, "Change and Motion," in *Cambridge History of Science: Volume 2*, 435.

35 David C. Lindberg and Katherine H. Tachau, "The Science of Light and Color, Seeing and Knowing," in *Cambridge History of Science: Volume 2*, 503–4.

36 A. Mark Smith, *From Sight to Light: The Passage from Ancient to Modern Optics* (Chicago: University Of Chicago Press, 2014), inside jacket synopsis. See Johannes Kepler and William H. Donahue, *Optics: Paralipomena to Witelo and Optical Part of Astronomy* (Santa Fe, NM: Green Lion Press, 2000). This was the overdue first translation of Kepler's Latin optical-astronomical work into any language.

37 Edward Grant, "Science and the Medieval University," in *Rebirth, Reform, and Resilience: Universities in Transition 1300–1700*, ed. James M. Kittelson and Pamela J. Transue (Columbus: Ohio State University Press, 1984).

38 Michael H. Shank, "Myth 2: That the Medieval Christian Church Suppressed the Growth of Science," in Ronald L. Numbers, ed. *Galileo Goes to Jail: And Other Myths About Science and Religion* (Cambridge, MA: Harvard University Press, 2009), 22.

39 A series of university curricular "condemnations" (targeting especially certain Aristotelian ideas) by a few local ecclesiastical authorities in the thirteenth century had very little effect on what actually was taught in most universities—even in the local settings in which the condemnations occurred. The very limited influence of these condemnations may have (ironically but modestly) stimulated the progress of science because they encouraged thinking outside the Aristotelian box.

40 John L. Heilbron, *The Sun in the Church: Cathedrals as Solar Observatories* (Cambridge, MA: Harvard University Press, 1999), 3.

41 Shank, "Myth 2," 21.

42 Tim O'Neill, "'The Dark Ages'—Popery, Periodisation, and Pejoratives," *History for Atheists*, November 19, 2016, https://historyforatheists.com/2016/11/the-dark-ages-popery-periodisation-and-pejoratives/.

43 Edward Grant, a leading historian of medieval science, suggested the label "Age of Reason" in *God and Reason in the Middle Ages* (Cambridge: Cambridge University Press, 2001), 30.

Chapter 3: Flat Earthers

1 Natalie Wolchover, "Are Flat-Earthers Being Serious?" *Live Science*, May 30, 2017, https://www.livescience.com/24310-flat-earth-belief.html. For coverage of the tiny minority flat-earth belief of the nineteenth and twentieth centuries, see Christine Garwood, *Flat Earth: The History of an Infamous Idea* (New York: Thomas Dunne Books, 2008).

2 https://twitter.com/neiltyson/status/692939759593865216.

3 https://genius.com/Tyson-flat-to-fact-lyrics.

4 Rodney Stark, *For the Glory of God: How Monotheism Led to Reformations, Science, Witch-Hunts, and the End of Slavery* (Princeton: Princeton University Press, 2003).

5 Tyson later tried to patch his Twitter historical flat-earth error but in so doing made yet another error. See Tim O'Neill, "The Great Myths 1: The Medieval Flat Earth," *History for Atheists*, July 1, 2016, https://historyforatheists.com/2016/06/the-great-myths-1-the-medieval-flat-earth. Meanwhile, B.o.B has not backed down in the face of Tyson and his ten million Twitter followers. In September 2017 he started a "Show BoB the Curve" GoFundMe campaign aimed at launching weather balloons and research satellites to confirm a flat earth. Even with 2.5 million Twitter followers, his campaign managed to raise less than $7,000 toward its $1 million goal in the first eleven months. Later in September 2017, Tyson tweeted back with a Photoshopped image of our moon with a flat-shaped Earth shadow partially eclipsing it. The caption: "A Lunar Eclipse flat-Earther's [sic] have never seen." See Yasmin Tayag, "Neil deGrasse Tyson Slams Flat Earth Theory with a Single Picture," *Inverse*, November 27, 2017, https://www.inverse.com/article/38783-neil-degrasse-tyson-flat-earth-conspiracy.

6   Draper, *History of the Conflict Between Religion and Science*, 157–59.

7   See also historian of science Lawrence Principe, "Transmuting History," *Isis* 98, no. 4 (2007): 786. About 70 percent of Principe's (mainly American) students over the past decade were taught in grade school that Columbus "proved the earth was round."

8   Thomas Smith and Donald M'Donald, *A Compendious System of Astronomy*, 6 vols., vol. 1, The Scientific Library (New York: D. M'Donald and J. Gillet, 1818). The first edition appeared in London in 1806.

9   Washington Irving, *A History of the Life and Voyages of Christopher Columbus*, 3 vols., vol. 1 (New York: G. and C. Carvill, 1828), 77.

10   Ibid., 77–78. Given that no official records of this consultation have survived, we must rely on accounts given years later, such as those of Columbus's son Fernando, who is reported to have remembered that some scholars, "basing their opinions more on the science of cosmography, said that the world was so large that it was not credible that three years time would be enough to reach the limit of the east." Irving may have used this source. Fernando recalled other arguments against the success of the proposed trip, but he did not say that anyone questioned earth's roundness, as Irving claimed. Fernando Colombo, *The History of the Life and Deeds of the Admiral Don Christopher Columbus: Attributed to His Son Fernando Colón*, ed. Ilaria Caraci Luzzana, trans. Geoffrey Symcox and Blair Sullivan (Turnhout: Brepols, 2004), 56.

11   Lesley B. Cormack, "Flat Earth or Round Sphere: Misconceptions of the Shape of the Earth and the Fifteenth-Century Transformation of the World," *Ecumene* 1, no. 4 (1994): 371. See also Lesley B. Cormack, "That Before Columbus Geographers and Other Educated People Thought the Earth Was Flat," in Numbers and Kampourakis, *Newton's Apple and Other Myths About Science*. In the conference for the authors preparing their contributions to this myth-busting anthology (which includes my piece on the Copernican demotion myth), I served as the commentator on Cormack's flat-myth paper.

12   Irving, *Life and Voyages of Christopher Columbus*, 78.

13   Rudolf Simek, *Heaven and Earth in the Middle Ages: The Physical World Before Columbus*, trans. Angela Hall (Rochester, NY: Boydell Press, 1997), 25.

14   Darley was a poet, novelist, literary critic, and mathematician whose career included writing for the *London Magazine* under the pseudonym John Lacy. John W. Cousin, "A Short Biographical Dictionary of English Literature," http://www.gutenberg.org/files/13240/13240-h/13240-h.htm.

15   George Darley, *Familiar Astronomy* (London: John Taylor, 1830), 25–27.

16   David P. Todd, *A New Astronomy* (New York: American Book Company, 1897), 77.

17   Forest Ray Moulton, *An Introduction to Astronomy* (New York: Macmillan, 1906), 93.

18   Edward Arthur Fath, *The Elements of Astronomy*, 4th ed. (New York: McGraw-Hill, 1944), 25.

19   D. Scott Birney, *Modern Astronomy* (Boston: Allyn and Bacon, 1969), 15. "Mankind, with few exceptions, seems always to have held the belief that the earth is approximately flat, and it was not until Magellan's ships actually sailed around it that the agelong belief was shattered and the rotundity of the earth generally accepted as a fact."

20   John D. Fix, *Astronomy: Journey to the Cosmic Frontier*, 6th ed. (New York: McGraw-Hill, 2011), 58.

21   This is where, in the first century, followers of Jesus first became known as "Christians," as recorded in Acts 11:26.

22   C. P. E. Nothaft, "Augustine and the Shape of the Earth: A Critique of Leo Ferrari," *Augustinian Studies* 42, no. 1 (2011): 37–38. For a stunning refutation of the latest attempt to argue *against* the dominance of the spherical-earth model before 1492, see Nothaft, "Zaccaria Lilio and the Shape of the Earth: A Brief Response to Allegro's 'Flat Earth Science,'" *History of Science* 55, no. 4 (2017): 490–98. Allegro declined an invitation to respond to Nothaft. That says much after you compare the two papers.

23 Cormack, "Flat Earth or Round Sphere," 366.

24 Nothaft, "Augustine and the Shape of the Earth," 39.

25 Jeffrey Burton Russell, *Inventing the Flat Earth: Columbus and Modern Historians* (New York: Praeger, 1991), 32.

26 Nothaft, "Augustine and the Shape of the Earth," 45.

27 Ibid., 40. Augustine did not investigate whether earth moved so as to make the sun appear to move. Virtually nobody did until the late Middle Ages. In more recent times the word *literal* (for many people) has come to mean something different in matters of biblical interpretation.

28 Ibid., 47–48.

29 Stephen C. McCluskey, "Natural Knowledge in the Early Middle Ages," in *Cambridge History of Science: Volume 2, Medieval Science*, ed. David C. Lindberg and Michael H. Shank (Cambridge: Cambridge University Press, 2013), 305.

30 Simek, *Heaven and Earth in the Middle Ages*, 31. Sacrobosco's *De sphaera* and its advanced companion, the *Theorica planetarum*, together constituted the leading astronomy text-book of medieval and early modern universities, going through numerous editions in manuscript and print. Simek further notes: "Even more elementary was another piece of evidence for the sphericity of the earth—and therefore also its seas—mentioned by John of Sacrobosco, which is accompanied in almost all the [medieval] handbooks by an illustration. If a ship leaves the shore, then the ship's hull soon disappears over the horizon whereas the mast-top remains visible for longer. The disappearance of this ship is therefore not caused by the increasing distance but by the curvature of the surface of the sea."

31 Jeffrey Burton Russell, "The Myth of the Flat Earth," http://www.asa3.org/ASA/top-ics/history/1997Russell.html. See also his fine monograph: Russell, *Inventing the Flat Earth*.

32 The Joint Caucus of Socially Engaged Philosophers and Historians of Science, http://jointcaucus.philsci.org, "was founded in 2012 to promote research, educational and public activities in history and philosophy of science that constructively engages mat-ters of social welfare." My public testimony regarding science-education standards is an example of such involvement.

33 Although medieval disputations were more focused on debating written texts about nature rather than extensive firsthand encounters with nature itself, even today under-graduate science majors acquire the vast majority of knowledge of nature by interacting with scientific texts. In laboratory course components, the professor and the laboratory manual largely guide students to see and interpret nature in certain ways. I have taught lab-based science for many years and have reflected on that experience as a philosopher-historian of science. Even graduate students doing original research have minds filled with "texts," whether acquired by reading or by listening to professors and other students. There are virtually no text-free encounters with nature in scientific practice or scientific pedagogy, whether medieval or modern. Of course, one can find many differences in how science is practiced and taught depending on the historic period, the particular field of science, and other factors, but there is also much continuity in the human condition in the face of nature. Making too sharp a distinction between medieval science and modern science is erroneous.

34 Edward Grant, *The Foundations of Modern Science in the Middle Ages: Their Religious, Institutional, and Intellectual Contexts* (Cambridge: Cambridge University Press, 1996), 41–42.

35 I first encountered this in James Hannam, *God's Philosophers: How the Medieval World Laid the Foundations of Modern Science* (London: Icon Books, 2009), 35.

36 Quoted in John Henry, *Knowledge Is Power: How Magic, the Government, and an Apoca-lyptic Vision Helped Francis Bacon to Create Modern Science* (Cambridge: Icon, 2003), 85.

37 Francis Bacon, *The New Organon and Related Writings*, ed. Fulton Anderson (New York:

Liberal Arts Press, 1960), 87, sect. 89: "Neither is it to be forgotten that in every age natural philosophy has had a troublesome and hard to deal with adversary—namely superstition and the blind and in moderate zeal of religion. For we see among the Greeks that those who first proposed to men's then uninitiated years the natural causes for thunder and for storms were thereupon found guilty of impiety. Nor was much more forbearance shown by some of the ancient fathers of the Christian church to those who on most convincing grounds (such as no one in his senses would now think of contradicting) maintained that the earth was round, and of consequence asserted the existence of the antipodes. Moreover, as things now are, to discourse of nature is made harder and more perilous by the summaries and systems of the schoolmen." See the debate over whether Bacon's apparently vigorous Protestantism was pious, conventional, or even deist or atheist camouflage in Steven Matthews, *Theology and Science in the Thought of Francis Bacon* (Burlington: Ashgate, 2008), and the review by Cameron Wybrow in *Early Science and Medicine*, 15 (2010): 302–4. Which of these interpretations is true would have a bearing on how we evaluate Bacon as a mythmaker.

Chapter 4: Burning Bruno

1   Giordano Bruno, *Opere di Giordano Bruno Nolano*, ed. Adolf Wagner (Leipzig: Weidmann, 1830).
2   Rachel Bishop, "Alien Enthusiast Mysteriously Disappears Leaving Behind Locked Code-Covered Room and Statue of 16th-Century Philosopher Who Predicted Extraterrestrial Life," *The Mirror*, April 6, 2017, http://www.mirror.co.uk/news/world-news/student-mysteriously-disappears-leaving-behind-10162430; Pollux, "The Mysterious Disappearance of Bruno Borges," *Skeptico Community*, June 1, 2017, http://www.skeptiko-forum.com/threads/the-mysterious-disappearance-of-bruno-borges.3850/.
3   Rio Branco, "Busquei o isolamento pra não ser atrapalhado pelo coletivo', diz Bruno Borges, sem dizer onde ficou," *Globo*, August 13, 2017, translation available at https://translate.google.com/translate?depth=1&hl=en&ie=UTF8&prev=_t&rurl=translate.google.com&sl=pt-BR&sp=nmt4&tl=en&u=http://g1.globo.com/ac/acre/noticia/busquei-o-isolamento-pra-nao-ser-atrapalhado-pelo-coletivo-diz-bruno-borges-sem-dizer-onde-ficou.ghtml. I found this Brazilian news story by means of George Noory, "'Missing' Occult Author Returns," *Coast to Coast AM*, August 22, 2017, https://www.coasttocoastam.com/article/missing-occult-author-returns/.
4   John William Draper, *History of the Conflict Between Religion and Science* (New York: D. Appleton, 1874), 180–81.
5   Ingrid D. Rowland, *Giordano Bruno: Philosopher/Heretic* (New York: Farrar, Straus and Giroux, 2008), 5. Though Rowland is a highly gifted storyteller, several reviewers have found her book plagued with errors: Brad S. Gregory, "Giordano Bruno Superstar," *Books and Culture* 18, no. 2 (2012):19–21; Owen Gingerich, "The Long Road to Infinity," *Wall Street Journal*, December 19, 2008, A15. Gingerich observes that "though her book is full of references to Bruno's philosophy and mathematics, they remain hard to follow: One is often left not knowing precisely what he was arguing about, beyond the existence of the infinite and the infinitesimal. This murkiness is partly Bruno's fault, but not entirely."
6   Camille Flammarion, *Popular Astronomy: A General Description of the Heavens*, trans. J. Ellard Gore (New York: D. Appleton, 1890).
7   Ibid., 254.
8   Herbert A. Howe, *Elements of Descriptive Astronomy: A Text-Book* (New York: Silver, Burdett, and Co., 1897), 310.
9   Cecilia Helena Payne-Gaposchkin, *Introduction to Astronomy* (New York: Prentice-Hall, 1954), 163.

10   Jole Shackelford, "Myth 7: That Giordano Bruno Was the First Martyr of Modern Science," in Ronald L. Numbers, ed. *Galileo Goes to Jail: And Other Myths About Science and Religion* (Cambridge, MA: Harvard University Press, 2009), 65.

11   Dionysius of Alexandria, "Against the Epicureans," in Alexander Roberts and James Donaldson, *Ante-Nicene Christian Library*, vol. 20: *The Writings of Gregory Thaumaturgus, Dionysius of Alexandria, and Archelaus* (1871), 171.

12   Basil invoked God's ability to make many "heavens" or "worlds" beyond the one conceived by Aristotle, who in turn, Basil noted, had argued against those who believed "there are infinite heavens and worlds." *Saint Basil: Exegetic Homilies*, The Fathers of the Church: vol. 46 (Washington, DC: Catholic University of America Press, 1963), Homily 3, 40.

13   Alberto A. Martinez, "Giordano Bruno and the Heresy of Many Worlds," *Annals of Science* 73, no. 4 (2016): 350.

14   Ibid., 351. Martinez reports that around AD 384 Philaster wrote: "Another heresy is to say that worlds are infinite and innumerable, following the asinine opinion of the philosophers, whereas Scriptures say that the world is one and it teaches us that it is one. This is also about the prophets' apocrypha (that is, secrets), who said that those who believe this are pagans; such as Democritus who asserted that many worlds exist, with which he proclaimed his wisdom, he stirred many souls to experience various doubtful errors." Around 472 Saint Augustine accepted Philaster's categorization of an infinite number of worlds as heretical when Augustine composed a list of eighty-eight heresies that included "that worlds are innumerable." But Augustine considered Epiphanius (who identified and refuted eighty heresies) as more learned than Philaster for such a task, as indicated in Augustine's letter to Quodvultdeus quoted in Nathaniel Lardner, *The Works of Nathaniel Lardner*, vol. 4 (London: William Ball, 1838), 385.

15   Bruce M. Metzger, *The Canon of the New Testament* (Oxford: Clarendon Press, 1987), 233.

16   Ibid. Metzger also observes that in *Liber de haeresibus* (Book of Heresies), Philaster "sweeps together an ill-digested assortment of comments compiled from Greek and Latin authors without much regard for logic or even internal consistency." Johann Lorenz Mosheim et al., *Institutes of Ecclesiastical History, Ancient and Modern*, 4th ed. (London: Longman, Roberts, and Green, 1863), 262, observes that *Liber de haeresibus* "enumerates more heresies than any of the other ancient works but no one considers it an accurate and able work. Philastrius was doubtless a pious and well-meaning man; but he was incompetent to the task that he undertook."

17   Martinez, "Giordano Bruno," 351.

18   This translation is from Martinez, "Giordano Bruno," 354.

19   John Carey, "Ireland and the Antipodes: The Heterodoxy of Virgil of Salzburg," *Speculum* 64, no. 1 (1989): 1. Carey's translation of "quod alius mundus, et alii homines sub terra sint, seu sol et luna" is: "there are another world and other men beneath the earth, or even the sun and moon." Carey notes that the phrase "seu sol et luna" has generally been rendered "or (another) sun and moon," leading to some confusion. Carey says that various scholars have claimed that this common translation distorts Virgilius's opinion, which was focused on the antipodes, not extrasolar planets.

20   Antipodes (from the Greek, meaning "opposite feet") is a term derived from the *opposite* orientation of the *feet* of any inhabitants (if any exist) of this inaccessible land conjectured to exist in earth's other hemisphere. But if those opposite feet belong only to animals, or if these conjectured inaccessible continents are completely uninhabited, then there ceases to be a theological dispute over the matter. Eventually Columbus did discover what became known as the New World. The Church welcomed many converts from this alien world and baptized them.

21   Pope Zacharias further declared that if Virgilius "is convicted by the summoned council of the Church for confessing this, he will be deprived of the honor of the priesthood."

Because Virgilius was not removed from the priesthood, but rather was made a bishop and even canonized, Martinez infers that Virgilius "defended himself convincingly." Perhaps so, but his defense probably related to inhabited antipodes, not ET.

22 Edward Grant, *The Foundations of Modern Science in the Middle Ages* (Cambridge: Cambridge University Press, 1996), 78.

23 Michael J. Crowe, *The Extraterrestrial Life Debate, 1750–1900* (New York: Dover, 1999), 6. This is a Dover reprint of the original 1986 book with Cambridge University Press.

24 James Hannam, *God's Philosophers: How the Medieval World Laid the Foundations of Modern Science* (London: Icon Books, 2009), 105, building on Grant, argues that these condemnations do not indicate that the Church was anti-science. "The limits imposed on natural philosophy served a dual purpose. While they did prevent it from impinging on theology, they also protected natural philosophers from those who wanted to see either activities further curtailed. Like a country with secure borders, philosophy was safe to develop in peace and without fear."

25 Michael J. Crowe, *The Extraterrestrial Life Debate, Antiquity to 1915: A Source Book* (Notre Dame: University of Notre Dame Press, 2008), 26.

26 Dennis R. Danielson, *The Book of the Cosmos* (Cambridge, MA: Basic Books, 2001), 92–101. Galileo later used similar arguments, which indicates his medieval indebtedness.

27 Crowe, *The Extraterrestrial Life Debate, Antiquity to 1915*, 27–34. Nicholas admitted that "of the inhabitants then of worlds other than our own we can know still less, having no standards by which to appraise them" (32).

28 Crowe, *The Extraterrestrial Life Debate, 1750–1900*, 8.

29 Crowe, *The Extraterrestrial Life Debate, Antiquity to 1915*, 27.

30 Crowe, *The Extraterrestrial Life Debate, Antiquity to 1915*; Crowe, *The Extraterrestrial Life Debate, 1750–1900*.

31 Martinez, "Giordano Bruno," 357.

32 *Corpus Juris Canonici* (1582), http://digital.library.ucla.edu/canonlaw/.

33 Martinez, "Giordano Bruno," 357–58.

34 Ibid., 358–59.

35 Ibid.

36 Miguel A. Granada, "Mersenne's Critique of Giordano Bruno's Conception of the Relation Between God and the Universe: A Reappraisal," *Perspectives on Science* 18, no. 1 (2010): 29–30.

37 Alberto A. Martinez, *Burned Alive: Giordano Bruno, Galileo, and the Inquisition* (London: Reaktion Books, 2018), chap. 3, writes: "Like Campanella, Mersenne just didn't know that belief in many worlds was officially a heresy." More likely they knew that belief in many worlds was *not* officially heretical, as I have shown. Martinez's case weakens each time he has to write off another prominent early modern character as ignorant about what was considered officially heretical.

38 Granada, "Mersenne's Critique of Giordano Bruno's Conception," 29–30.

39 Ibid., 32, 39–40.

40 Ibid., 33.

41 Ibid., 36.

42 Alberto A. Martinez, "Ten Censured Propositions in Giordano Bruno's Books," *Bruniana Campanelliana Brunianae Campanelliana* 22, no. 2 (2016): 411–26. Similarly, note Martinez's conclusion to his chapter on Bruno in his book on Pythagorean heresies: "Many of Bruno's main philosophical and religious heresies were directly rooted in the ancient and *pagan Pythagorean religion* and its interpretations in the Renaissance. Even seemingly scientific notions, such as the motion of the Earth, were connected to Pythagorean ideas such as that the Earth has a soul. The idea that there are many worlds like ours was connected to the Pythagorean belief that the beings living in such worlds can embody the souls of persons who were once alive on Earth." Alberto A. Martinez, *Pythagoras, Bruno, Galileo: The Pagan Heresies of the Copernicans* (Cambridge, MA: Saltshadow Castle, 2014), 186.

43   Maurice A. Finocchiaro, "Philosophy Versus Religion and Science Versus Religion: The Trials of Bruno and Galileo," in *Giordano Bruno: Philosopher of the Renaissance*, ed. Hilary Gatti (Aldershot: Ashgate, 2002). Martinez, agreeing with Finocchiaro, says that "the conflict between Bruno and the Inquisition can hardly be described as a conflict between science and religion." Martinez, "Giordano Bruno," 374.

44   Finocchiaro, "Philosophy Versus Religion and Science Versus Religion," 65.

45   Ibid. Finocchiaro explains: "Here the logic of the Inquisition procedure was that his obstinacy in not retracting the theses to which he had confessed [such as many worlds and transmigration of souls from body to body] rendered him guilty of the other opinions [such as denying Christ's deity] of which he had been accused but which had not been otherwise proved [to be actually held by him]."

46   Ibid, 56.

47   Ibid., 61. Martinez, "Ten Censured Propositions," 415–16, notes that the Council of Sens (1141) declared as heretical Peter Abelard's advocacy of the Holy Spirit as identical to the world soul—a ruling confirmed by Pope Innocent II. During his trial Bruno said he did not understand the Trinity or "the Holy Spirit as a third person," except "by following the Pythagorean way" as soul of the cosmos. Another heresy that the Inquisition identified in Bruno's books was the claim that every human soul is derived from the world soul.

48   Martinez, "Ten Censured Propositions," 420.

49   Finocchiaro, "Philosophy Versus Religion and Science Versus Religion," 64.

50   William R. Shea and Mariano Artigas, *Galileo Observed: Science and the Politics of Belief* (Sagamore Beach, MA: Science History Publications, 2006), 135.

51   Rowland, *Giordano Bruno*, chap. 8.

52   Ibid., 221.

53   Hilary Gatti, *Essays on Giordano Bruno* (Princeton: Princeton University Press, 2011), 318–19.

54   A complete English translation of this eyewitness account is at https://historyforatheists.com/2017/05/giordano-bruno-gaspar-schoppes-account-of-his-condemnation.

55   Martinez comes close to claiming just this in the introduction to *Burned Alive*. He praises Bruno for reaching modern-looking conclusions about exoplanets, star motion through different depths in space, and more. Martinez asks: "Who was this man who was right about so many things? Why has he been so neglected and disdained in the history of astronomy?"

56   Crowe, *The Extraterrestrial Life Debate, Antiquity to 1915*, 46.

57   Frances Amelia Yates, *Giordano Bruno and the Hermetic Tradition* (London: Routledge and Kegan Paul, 1964). Yates traced Bruno's mystical and animistic religious thought back to the Hermetic magical tradition. She overstated her thesis, as subsequent scholarship has shown. I do not rely on her work in this essay. The parts of her book that have withstood scrutiny would probably strengthen the case I am making here.

58   Hannam, *God's Philosophers*, 307. Although even clearly reputable scientists such as Kepler also speculated about souls in celestial bodies, they placed much more emphasis on mathematically precise theory supported by observations, something mostly foreign to Bruno's mentality.

59   Shea and Artigas, *Galileo Observed*, 129; David Wootton, *The Invention of Science: A New History of the Scientific Revolution* (New York: HarperCollins, 2015), 30; Hilary Gatti, *Giordano Bruno and Renaissance Science* (Ithaca, NY: Cornell University Press, 1999), 84.

60   Giordano Bruno, *The Ash Wednesday Supper*, trans. Stanley L. Jaki (Paris: Mouton, 1975), 56–67; also Crowe, *The Extraterrestrial Life Debate, Antiquity to 1915*, 49–50.

61   Gatti, *Giordano Bruno and Renaissance Science*, 123, notes that later Bruno declared that Copernicus had put the moon on an epicycle centered on earth, but his own thermodynamic theory could not explain how this would work.

62   Ernan McMullin, "Bruno and Copernicus," *Isis* 78, no. 1 (1987): 55–74. Robert West-

man thinks this error can be rationalized, in part, by taking Bruno to have imposed on Copernican astronomy an ancient "Pythagorean" cosmology, according to which the earth and an invisible counter-earth revolve around a cosmically "Central Fire." On this account, Bruno substituted the moon for the counter-earth and the sun for the Central Fire. But this still results in a theoretical moon that does not behave like the moon we see in the sky with its signature monthly phases. Robert S. Westman, *The Copernican Question: Prognostication, Skepticism, and Celestial Order* (Berkeley: University of California Press, 2011), 303.

63 Westman, *The Copernican Question*, 301.

64 He had chiefly aimed, McMullin notes, to "help others ascend, as he claims to have done, to an intuitive knowledge of the divine. But the steps are still the steps of argument, however allegorical. We are unlikely to find them persuasive, needless to say." McMullin, "Bruno and Copernicus," 70.

65 Paul-Henri Michel, *The Cosmology of Giordano Bruno*, trans. R. E. W. Maddison (Ithaca, NY: Cornell University Press, 1973), 296–302, as cited in McMullin, "Bruno and Copernicus," 73. McMullin responds to Michel: "Perhaps he has in mind the rather tenuous notion of 'prophetic intuition' that Bruno himself attributes to Aristotle and Copernicus, a gift of getting things right for the wrong reasons."

66 Gatti, *Giordano Bruno and Renaissance Science*, 111–12. In an attempt to summarize Bruno's argument, she writes that "an all-powerful God can, as a logical necessity, express himself only in terms of *potentia absoluta,* and therefore his entire creation will partake of the infinite."

Chapter 5: Gagging Galileo

1 This is widely considered the first "New Atheist" book: Sam Harris, *The End of Faith: Religion, Terror, and the Future of Reason* (New York; London: W. W. Norton, 2004). For this assessment see Massimo Pigliucci, "New Atheism and the Scientistic Turn in the Atheism Movement," *Midwest Studies in Philosophy* 37, no. 1 (2013): 142.

2 J. L. Heilbron, *Galileo* (Oxford: Oxford University Press, 2010), 150.

3 Ibid., 153.

4 Additional strengths at the time included: the Tychonic system expected a *lack* of stellar parallax and it better made sense of the presence of detectable stellar widths. Both of these strengths were later erased by better technology and new scientific discoveries. Even so, at the time, the Tychonic system was very well supported by evidence. See Christopher M. Graney, *Setting Aside All Authority: Giovanni Battista Riccioli and the Science Against Copernicus in the Age of Galileo* (Notre Dame: University of Notre Dame Press, 2015).

5 William R. Shea and Mariano Artigas, *Galileo in Rome: The Rise and Fall of a Troublesome Genius* (Oxford: Oxford University Press, 2003), 30–43.

6 Galileo Galilei, *Selected Writings: Galileo Galilei,* trans. William R. Shea and Mark Davie (Oxford: Oxford University Press, 2012), 82–83.

7 Ibid., 80–81.

8 Ibid., 70.

9 Ibid., 82.

10 Near the end of his Christina letter, Galileo seems to argue that the Bible actually supports one element of his updated version of the Copernican system. This argument, if serious, is inconsistent with Galileo's earlier principle according to which the Bible teaches *no* cosmology whatsoever. Galileo cited Joshua's long day, when Joshua commanded the sun to stand still and the "sun stopped in the midst of heaven" (Joshua 10:13). The astronomer argued that this meant the sun stopped spinning on its axis in the center (midst) of the cosmos (heaven), which stopped all other planetary motion. This one

miracle would lengthen that particular day on earth to enable the Israelites to extend a winning battle with their enemy. But in making this case, Galileo might have been "reducing to absurdity" the counterclaim that the Bible supports Ptolemaic cosmology. In effect, he might have been saying, if you want to play the illegitimate game of using the Bible to support a particular cosmology, well I can do that too—see how ridiculous it is! See "Letter to the Grand Duchess Christina," in Maurice Finocchiaro, *The Essential Galileo* (Indianapolis: Hackett, 2008), 140–44, or the more recent translation of Shea and Davie, *Selected Writings: Galileo Galilei*, 89–93. Alternatively, Galileo might have been arguing that once you settle by science whether the sun or the earth moves, then you should interpret the Bible in a way that is consistent with such truths known through science. This interpretation is consistent with what Galileo said in his Christina letter about Joshua's long day a few pages before he dealt with the topic in detail: "But in any case, I will show below that it is necessary to gloss and interpret the meaning of the text of the book of Joshua regardless of the view we take of the structure of the universe" (p. 84 in the Shea-Davie translation).

11   Annibale Fantoli, "The Disputed Injunction and Its Role in Galileo's Trial," in *The Church and Galileo*, ed. Ernan McMullin (Notre Dame: University of Notre Dame Press, 2005), 124–26.

12   "Bellarmine's Letter to Foscarini," in Finocchiaro, *The Essential Galileo*, 147.

13   Ibid.

14   Some historians of science have concluded that Bellarmine intended his letter to convey a quite different message from what I present in this chapter. Ernan McMullin is one of them. McMullin argues that Bellarmine's "very great doubts" that a "true demonstration" of heliocentrism would ever be achieved really amounted to the *denial* of any possible successful scientific future for Copernicanism. So Bellarmine's "very great doubts," according to McMullin, merely expressed "the innate courtesy for which Bellarmine was famous." To support this interpretation, McMullin notes that Bellarmine "went on in the remainder of the letter to list several reasons why such a proof would not be forthcoming."

McMullin claims that Bellarmine suggested that all astronomical theories were merely useful ways of accounting for appearances, without explaining the actual physical motion of celestial objects. Actually, Bellarmine addressed only something similar to this topic, and he mentioned it before rather than after (as McMullin says) citing his "very great doubts" about a future proof of Copernicanism. So it was not part of a *subsequent* "list" of "several reasons why such a proof would not be forthcoming," as McMullin argues. Furthermore, even though Bellarmine mentioned earlier in his letter the precedent for something like this non-truth-producing (instrumentalist) view of astronomy, he acknowledged in the same passage that some people wanted to "affirm that in reality the sun is the center of the world." He said this without explicitly denying that such people could be legitimate, noninstrumentalist, realist astronomers. *Later* in the letter, as noted, Bellarmine acknowledged that if there were a "true demonstration" of heliocentrism as physical reality, the Bible would have to be reinterpreted accordingly.

A passage from a Bellarmine lecture further undermines McMullin's claim that Bellarmine conceived of astronomy exclusively in a non-truth-producing (instrumentalist) manner. As a twenty-nine-year-old professor at the University of Louvain, Bellarmine examined Thomas Aquinas's discussion of how the stars appear to move around the earth. Noting different explanations for apparent star motion, including the traditional Ptolemaic one and an explanation based on "the movement of the earth," Bellarmine said that "if one ascertained with evidence" that one of these views was correct, then "one would have to consider a way of interpreting the Scriptures which would put them in agreement with the ascertained truth: for it is certain that the true meaning of Scripture cannot be in contrast with any other truth." So Bellarmine, despite McMullin's interpretation otherwise, thought it was possible to discover truth about physical reality in astronomy.

Moreover, McMullin claims that Bellarmine *showed* in his letter that he was not really open to a possible Copernican proof by asserting that any attentive person already "clearly experiences that the earth stands still and that the eye is not in error when it judges that the sun moves." This was part of Bellarmine's response to a relativity-of-motion argument that had been discussed in medieval and early modern times, including by Galileo. According to this argument, when a ship leaves shore, it appears (from the perspective of those on the ship) that the shore is moving away from the stationary ship. Perhaps we experience earth as apparently motionless for similar reasons. When talking about the natural world, the Bible could legitimately accommodate its language to the common experience of how things appear to the human eye: earth seems to be at rest and the sun appears to revolve around us. So Bellarmine was responding to this relativity-of-motion argument and its implications for biblical interpretation.

McMullin correctly notes that Bellarmine's attempted argument here is guilty of "begging the question"—assuming the very thing that one is trying to prove (that the sun, not the earth, does the moving). Bellarmine, a sophisticated theologian trained in Aristotelian logic, probably knew just how desperate and question begging his "answer" to the relativity-of-motion argument really was. Jules Speller, in criticizing McMullin, writes, "But I think that, because of its obvious flaws, 'Bellarmine's answer' looks more like the cardinal's desperate attempt to justify his doubts about the possibility of a demonstration of Copernicus rather than like an indication that he deems such a demonstration altogether impossible." See Jules Speller, *Galileo's Inquisition Trial Revisited* (Frankfurt: Peter Lang, 2008), 70.

So Bellarmine probably was a moderately open-minded theologian about a possible future scientific demonstration of Copernicanism. Ironically, Galileo failed to produce the very physical proof of Copernicanism that he thought he had discovered—his theory of the rhythmic tidal movement of the ocean.

J. L. Heilbron's reputable Galileo biography also does not support McMullin's interpretation of Bellarmine's April 1615 letter. His take is consistent with mine and Speller's. Furthermore, Heilbron concludes that Bellarmine was "quite prepared to discover things in the heavens unsuspected by Aristotle" (Heilbron, *Galileo*, 213). We have no reason to believe that Bellarmine contested the (largely affirming) report that he requested and received from his colleagues at the Roman College about Galileo's telescopic discoveries. So the Inquisitor was not a science hater. But he and his Inquisition colleagues did restrict Galileo's intellectual freedom.

What if McMullin is right about the Bellarmine letter, and Heilbron, Speller, and I are wrong? That would change how Bellarmine fits into my account, but it would not significantly alter the overall conclusion: the Galileo affair does not show an inevitable conflict between science and Christianity. Although Bellarmine was the leading inquisitor, many other voices represented diverse opinions in Galileo's encounter with the Church.

15   Graney, *Setting Aside All Authority*.
16   Maurice A. Finocchiaro, "That Galileo Was Imprisoned and Tortured for Advocating Copernicanism," in Ronald L. Numbers, ed. *Galileo Goes to Jail: And Other Myths About Science and Religion* (Cambridge, MA: Harvard University Press, 2009), 69.
17   "Decree of the Index," in Finocchiaro, *Essential Galileo*, 177.
18   Annibale Fantoli, *Galileo: For Copernicanism and for the Church* (Vatican City: Vatican Observatory Publications, 2003), 454.
19   Ibid. I corrected a typo in this quotation by noting its correct translation in Shea and Artigas, *Galileo in Rome*, 134. The contextual cues in brackets are Fantoli's. Galileo learned of this conversation from a March 16, 1630, letter from his disciple Castelli. Shea and Artigas also note: "What Urban VIII said to Campanella in 1630 is exactly what he had told Cardinal Zollern in 1624. Copernicanism was not a heresy, to be sure, but it went against the apparent fact that the Earth is at rest at the center of the world, something virtually all biblical scholars took for granted."

20   Finocchiaro, "That Galileo Was Imprisoned and Tortured for Advocating Copernican-ism," 70.

21   Thomas F. Mayer and Project Muse, *The Roman Inquisition: Trying Galileo* (Philadelphia: University of Pennsylvania Press, 2015), 217–18. Mayer observes: "Urban's penchant for increasingly autocratic behavior, including his housecleaning of the papal administra-tion beginning in mid-1632, and blithe disregard for the law, in which his brother Anto-nio almost matched him, manifested in the cases of Orazio Morandi, Fra Innocenzo, Tommaso Campanella, and Giacinto Centini, should not be downplayed."

22   Shea and Artigas, *Galileo in Rome*, 176.

23   Finocchiaro, "That Galileo Was Imprisoned and Tortured for Advocating Copernican-ism," 7.

24   Shea and Artigas, *Galileo in Rome*, 186.

25   Finocchiaro, "That Galileo Was Imprisoned and Tortured for Advocating Copernican-ism," 71.

26   Ibid.

27   Ibid.

28   Ibid., 73, and 251n7.

29   Ibid., 76–78.

30   Maurice A. Finocchiaro, *Retrying Galileo, 1633–1992* (Berkeley: University of Califor-nia Press, 2005).

31   John Ewing, *A Plain Elementary and Practical System of Natural Experimental Philosophy: Including Astronomy and Chronology* (Philadelphia: Hopkins and Earle, 1809), 2.

32   C. S. Lewis, "On the Reading of Old Books," in *God in the Dock: Essays on Theology and Ethics*, ed. Walter Hooper (Grand Rapids: Eerdmans, 1970), 200–207.

33   William Phillips, *Eight Familiar Lectures on Astronomy Intended as an Introduction to the Science* (New York: James Eastburn, 1818), 8–9.

34   John Lauris Blake, *First Book in Astronomy Adapted to the Use of Common Schools* (Boston: Lincoln and Edmands, 1831), 11.

35   George G. Carey, *Astronomy as It Is Known at the Present Day* (London: William Cole, 1824), 163.

36   Ormsby M. Mitchel, *Popular Astronomy* (New York: Phinney, Blakeman, and Mason, 1860), 142. The same passage, on the same page, persists in the final (seventh) edition of the book, which was printed almost a dozen times by the same publisher from the mid-1860s up through the mid-1870s. (Mitchel died in 1862.)

37   Edmund Beckett Denison and Pliny Earle Chase, *Astronomy Without Mathematics* (New York: Putnam, 1871), 35.

38   Joel Dorman Steele, *A Fourteen Weeks Course in Descriptive Astronomy* (New York: A. S. Barnes and Co., 1869), 31.

39   Galileo to Kepler, August 19, 1610, as cited in Fantoli, *Galileo*, 92.

40   Heilbron, *Galileo*, 372.

41   Fantoli, *Galileo*, 101–2.

42   Johannes Kepler, *Epitome of Copernican Astronomy*, trans. Charles Glenn Wallis, Great Books of the Western World, vol. 16 (Chicago: Encyclopedia Britannica, 1952), 845–48.

43   Finocchiaro, *Essential Galileo*, 183.

44   Maurice A. Finocchiaro, *The Routledge Guidebook to Galileo's Dialogue* (New York: Rout-ledge, 2014), sec. 8.3. In the last part of this section, Finocchiaro argues that "although Galileo does not discuss Tycho's alternative explicitly, he does so implicitly, and hence he is not really neglecting it. One reason stems from the fact that the Tychonic theory does, after all, share a crucial common element with the Ptolemaic system; that is, both hold the earth to be motionless at the center of the universe. That is, in both systems, the diurnal motion belongs to the whole universe except the earth, and the annual motion belongs to the sun. Therefore, all the Galilean arguments for the earth's motion and against the geostatic, geocentric thesis undermine Tycho's as well as Ptolemy's world

view. This is the case, for example, with the sunspot argument, the tidal argument, and the argument from the law of revolution." Despite all this, Galileo still gives the impression that his arguments were stronger than they actually were, given what was known at the time. He could have corrected some of this unfairness by engaging more with the Tychonic system. See Graney, *Setting Aside All Authority*.

45   We know that many of the textbook authors who perpetuated the Galileo myth were pious Protestants. For example, Steele, a very active Methodist, had a reputation for both piety and academic rigor: Anna Campbell Palmer, *Joel Dorman Steele, Teacher and Author* (New York: A. S. Barnes and Co., 1900). Mitchel, a devout Presbyterian, delivered lectures affirming the harmony of the Bible with science: Ormsby M. Mitchel, *The Astronomy of the Bible* (New York: Blakeman and Mason, 1863); Philip S. Shoemaker, "Stellar Impact: Ormsby Macknight Mitchel and Astronomy in Antebellum America," Dissertation (University of Wisconsin–Madison, 1991), 196–99. Chase, a Quaker, argued that Christianity and philosophy (including science) were in harmony, and that Christian belief was grounded in evidence: Pliny Earle Chase, "The Philosophy of Christianity," *Proceedings of the American Philosophical Society* 18, no. 103 (1879). Edmund Beckett Denison, designer of the clock mechanism behind the face of the British Parliament's famed Big Ben (1859), was an Anglican lawyer known for polemics within Anglican controversies such as the one in the 1890s over the remarriage of divorced persons (he argued for granting such remarriages). Protestant polemics against Catholics in regard to science and reason abounded in early nineteenth-century America, as documented by Lily A. Santoro, "The Science of God's Creation: Popular Science and Christianity in the Early Republic," Dissertation (University of Delaware, 2011), 22–24.

Chapter 6: Copernican Demotion

1   Biola University, "Does God Exist? William Lane Craig vs. Christopher Hitchens: Full Debate," April 4, 2009, https://www.youtube.com/watch?v=0tYm41hb48o.

2   Dennis Danielson, *The Book of the Cosmos* (New York: Basic Books, 2001), 106. This collection of sources includes excerpts of Copernicus's *On the Revolutions* (cited here from the seventh paragraph of the preface) and works by other theorists cited below.

3   C. S. Lewis, *The Discarded Image* (Cambridge: Cambridge University Press, 1964), 58.

4   Danielson, *The Book of the Cosmos*, 150.

5   There are rare cases in the early seventeenth century in which Copernican astronomy was seen as a demotion in conjunction with ET's existence. For example, Galileo discusses some objections to his celestial discoveries that some professors at the University of Perugia formulated. This is found in a long letter from Galileo to his Roman friend Piero Dini, May 21, 1611, *Opere di Galileo*, vol. XI, 105–16, as cited in William R. Shea and Mariano Artigas, *Galileo in Rome: The Rise and Fall of a Troublesome Genius* (Oxford: Oxford University Press, 2003), 120n208, 130. Shea and Artigas summarize Galileo's account of objections to earth as a planet: "If the Earth travelled around the Sun, it ceased to be at the center of the world and lost its distinctiveness. Changing its location entailed changing its nature. It was no longer unique but just one of several planets. A number of questions then raised their ugly heads, such as: Are there intelligent beings on other planets? And if so, how are we to understand the meaning of Original Sin, the incarnation, and the whole of redemption?" Pope Urban VIII, as a cardinal in 1615, mentioned some of these issues as well. But Shea and Artigas do not cite the reference for this.

6   Bernard le Bovier de Fontenelle, *A Plurality of Worlds*, trans. John Glanvill (London: R. Bentley and S. Magnes, 1688), 21.

7   Dennis Danielson, "The Great Copernican Cliché," *American Journal of Physics* 69 (2001): 1029–35.

8   Ibid., 1033.

9   Horatio N. Robinson, *A Treatise on Astronomy* (Albany: Erastus H. Pease, 1849), 103.

10 Dennis Danielson and Christopher M. Graney, "The Case Against Copernicus," *Scientific American* 310 (2013): 72–77.

11 Steven J. Dick, *The Biological Universe: The Twentieth-Century Extraterrestrial Life Debate and the Limits of Science* (Cambridge; New York: Cambridge University Press, 1996), 406–8; John Edward Westfall and William Sheehan, *Celestial Shadows: Eclipses, Transits, ad Occultations* (New York: Springer, 2015), 429; Julie Dobrow, "The Star-Crossed Astronomer," *Amherst Magazine*, July 28, 2017, https://www.amherst.edu/amherst-story/magazine/issues/2017-summer/the-star-crossed-astronomer.

12 David P. Todd, *A New Astronomy* (New York: American Book Company, 1897), 97.

13 Thomas R. Williams, "Abbot, Charles Greeley," in *The Biographical Encyclopedia of Astronomers*, ed. Thomas A. Hockey et al. (New York: Springer, 2014).

14 C. G. Abbot, *The Earth and the Stars* (New York: D. Van Nostrand, 1925), 8.

15 Danielson, *The Book of the Cosmos*, 150.

16 C. G. Abbot, "Religion and Man's Origin" (September 1926), printed pamphlet, Smithsonian Institution archive, box 199, http://siarchives.si.edu/collections/siris_arc_217205.

17 C. S. Lewis, *Miracles: A Preliminary Study* (San Francisco: Harper, 2001). This first edition of this book appeared in 1947.

18 Abbot, "Religion and Man's Origin."

19 Harlow Shapley, "The Scale of the Universe: Part I," *Bulletin of the National Research Council* 2, no. 11 (1921): 171.

20 Cecilia Helena Payne-Gaposchkin, *Introduction to Astronomy* (New York: Prentice-Hall, 1954), 2.

21 JoAnn Palmeri, "An Astronomer Beyond the Observatory: Harlow Shapley as Prophet of Science," PhD Dissertation (University of Oklahoma, 2000), 60, 69–72. Palmeri discovered that Shapley confessed to Harvard colleague Julian Coolidge: "I am willing to grant that astronomers should keep off of questions of ethics and esthetics and theologies and such things, but why do what we ought? Anyway, Kirsopp Lake says that professional philosophers have failed to advance philosophy, and that the best theology of the present time is science." Shapley added, "I am not sure if I am interested in astronomy for astronomy's sake, anyway." Shapley to J. Coolidge, December 12, 1923, UAV 630.22, box 4, as cited in Palmeri, "An Astronomer Beyond the Observatory," 178.

22 Hermann Bondi, *Cosmology* (Cambridge: Cambridge University Press, 1952), 13.

23 Owen Gingerich, *God's Universe* (Cambridge: Belknap Press of Harvard University Press, 2006), 14–15.

24 Stanley P. Wyatt, *Principles of Astronomy* (Boston: Allyn and Bacon, 1964), 152.

25 Ibid., 536–37.

26 William Lane Craig and J. P. Moreland, eds., *The Blackwell Companion to Natural Theology* (West Sussex, UK: Wiley-Blackwell, 2009), chap. 3; Geraint F. Lewis and Luke A. Barnes, *A Fortunate Universe: Life in a Finely Tuned Cosmos* (Cambridge: Cambridge University Press, 2016). The latter authors take a different approach (Lewis is an atheist and Barnes is a theist). Barnes affirms and critiques various aspects of Craig's work in *Letters to Nature*, https://letterstonature.wordpress.com/?s=+William+Lane+Craig.

27 Stephen E. Schneider and Thomas T. Arny, *Pathways to Astronomy*, 5th ed. (New York: McGraw-Hill, 2018), 629.

28 Neil deGrasse Tyson, foreword to Jeffrey O. Bennett et al., *The Cosmic Perspective*, 8th ed. (Boston: Pearson, 2017), xxviii.

29 Ibid.

30 Elaine Howard Ecklund, *Science vs. Religion: What Scientists Really Think* (Oxford: Oxford University Press, 2010), 57–58.

31 In the 2014 *Cosmos* TV series that Tyson hosted, and that President Obama endorsed in

its first episode, Tyson declared, "Accepting our kinship with all life on earth is not only solid science, in my view, it's also a soaring spiritual experience." *Cosmos: A Spacetime Odyssey*, www.cosmosontv.com, ep. 2, 37:00. Many teachers will probably use this series for science instruction over the next decades. We will analyze this series in chapter 9.

32   Danielson, "Copernican Cliché," 1033–34.

33   Eric Chaisson and Steve McMillan, *Astronomy: A Beginner's Guide to the Universe*, 8th ed. (Boston: Pearson, 2017), 26. This textbook is a scaled-down version of their full-length textbook *Astronomy Today* (same publisher). For myth statistics, I count only their full-version textbook.

34   Guillermo Gonzalez, Donald Brownlee, and Peter Ward, "The Galactic Habitable Zone: Galactic Chemical Evolution," *Icarus* 152 (2001): 185–200; David Waltham, *Lucky Planet* (London: Icon, 2014); John Gribbin, *Alone in the Universe* (Hoboken, NJ: Wiley, 2011); Peter Ward and Donald Brownlee, *Rare Earth* (New York: Copernicus, 2000).

35   Mark Lupisella, "Cosmocultural Evolution: The Coevolution of Culture and Cosmos and the Creation of Cosmic Value," in *Cosmos and Culture: Cultural Evolution in a Cosmic Context*, ed. Steven Dick and Mark Lupisella (Washington, DC: NASA, 2009), 343.

36   Gregory Allen Schrempp, *The Ancient Mythology of Modern Science: A Mythologist Looks (Seriously) at Popular Science Writing* (Montreal: McGill-Queen's University Press, 2012), 223.

37   Bennett et al., *The Cosmic Perspective*, 371.

38   Michael A. Seeds and Dana E. Backman, *Foundations of Astronomy*, 13th ed. (Boston: Cengage Learning, 2017), 626.

Chapter 7: Extraterrestrial Enlightenment

1   Paul Davies in *Science and Wonders: Conversations About Science and Belief*, ed. Russell Stannard, BBC Radio 4 Series (London: Faber, 1996), 73.

2   Frank Drake interview, "First Contact," SETI Institute, June 30, 2010, https://www.youtube.com/watch?v=zmfC51FstIg.

3   Frank D. Drake and Dava Sobel, *Is Anyone Out There? The Scientific Search for Extraterrestrial Intelligence* (New York: Delacorte Press, 1992), 160.

4   The younger Davies was less skeptical of godlike aliens. Davies, *Are We Alone? Philosophical Implications of the Discovery of Extraterrestrial Life* (Basic Books, 1995), 48: "It may be that, for us, the super advanced aliens would appear as gods."

5   Paul Davies, *The Eerie Silence: Renewing Our Search for Alien Intelligence* (Boston: Mariner Books, 2011), 151.

6   Because of the thermodynamic cosmic rule that will eventually produce the heat death of the universe, Drake's notion of salvation by ET would be temporary at best. Cosmic heat death means no more biological life or functioning computers.

7   Michael Shermer, *Skeptic: Viewing the World with a Rational Eye* (New York: Henry Holt, 2016), 125; "Shermer's Last Law," January 2002, http://www.michaelshermer.com/2002/01/shermers-last-law/.

8   Richard Dawkins, *The God Delusion* (Boston: Houghton Mifflin Co., 2006), 72.

9   Davies, *Eerie Silence*, 33.

10   Stephen Webb, *If the Universe Is Teeming with Aliens... Where Is Everybody?: Seventy-Five Solutions to the Fermi Paradox and the Problem of Extraterrestrial Life* (Cham, Switzerland: Springer, 2015), 78–123. Webb also analyzes imaginary colonization scenarios that would reduce the length of interstellar travel as ET civilizations and their robots spread throughout the cosmos. He notes that science fiction authors "have contributed at least as much to the debate as professional scientists," ix.

11   Ibid., 331.

12   Howard Smith, "Alone in the Universe," *Zygon* 51, no. 2 (2016): 512.
13   A light-year is, of course, the distance light travels in one year. Over one hundred generations, at twenty-five years per generation, light will travel 2,500 light-years. But that distance represents the upper limit for a *one-way* communication. For a communication to be sent *and* for the response to reach the originator of that communication within one hundred generations, the maximum distance between the two parties must be only half that 2,500 light-years—that is, 1,250 light-years *each way*. The equation looks like this: 100 x 25 / 2 = 1,250.
14   Smith, "Alone in the Universe," 504, estimated thirty million stars, but his assumptions probably inflated this outcome. Astronomer Guillermo Gonzalez, in personal email of September 16, 2018, used by permission, writes about Smith's estimate: "This is probably an overestimate. I get about 26 million assuming a constant star density equal to the local value with a sphere centered on the sun. But, the galactic disk is flattened on that scale, and the star density drops rapidly with vertical distance from the mid-plane. Probably 20 million stars would be a better estimate." My argument stands regardless of whether there are twenty million, thirty million, or even (say) a hundred million stars in this local region of the cosmos.
15   Guillermo Gonzalez, "Setting the Stage for Habitable Planets," *Life* 4, no. 1 (2014): 35–65. Open access: http://www.ncbi.nlm.nih.gov/pmc/articles/PMC4187148/.
16   Ibid.
17   Smith, "Alone in the Universe," 501.
18   Martin Rees, "Very Alien Intelligence," *New Scientist* 227, no. 3032 (2015): 22–23.
19   Webb, *If the Universe Is Teeming with Aliens*, 335.
20   Clément Vidal, "A Multidimensional Impact Model for the Discovery of Extraterrestrial Life," in *The Impact of Discovering Life Beyond Earth*, ed. Steven J. Dick (Cambridge: Cambridge University Press, 2015), 64.
21   Susan Schneider, "Alien Minds," in *The Impact of Discovering Life Beyond Earth*, 190.
22   Christopher Partridge, *The Occult World* (London: Routledge, 2016). The introduction to this scholarly anthology provides a useful overview of the occult in history and contemporary culture. The current use of the term *occult* (weird magical and spiritual phenomena) traces back to the nineteenth century. There is, however, a much longer history during which science and magic (occult) were closely related at various times, but then were distinguished more carefully. *Occult* means "hidden," and people debated whether specific occult phenomena were natural or supernatural.
23   Steven J. Dick, *The Biological Universe: The Twentieth-Century Extraterrestrial Life Debate and the Limits of Science* (Cambridge: Cambridge University Press, 1996), 250–51.
24   Ibid., 526.
25   Do not confuse this speculative technological singularity with the well-supported cosmological singularity that marks the beginning of the universe in today's standard cosmology.
26   Ray Kurzweil, *The Singularity Is Near: When Humans Transcend Biology* (New York: Viking Books, 2005), 5.
27   Ibid., 29.
28   Ibid., 375.
29   James Earl Adams III, "Which 150 Supplements Does Ray Kurzweil Take Daily?," *Quora*, April 25, 2016, https://www.quora.com/Which-150-supplements-does-Ray-Kurzweil-take-daily.
30   Murray Shanahan, *The Technological Singularity* (Cambridge: MIT Press, 2015), 91, 160.
31   Toby Walsh, "The Singularity May Never Be Near," Cornell University Library digital archive, February 20, 2016, https://arxiv.org/abs/1602.06462v1, 3.
32   Margaret A. Boden, *AI: Its Nature and Future* (Oxford: Oxford University Press, 2016), 56 and 153.
33   The frame problem concerns how to represent in formal logic the effects of actions with-

out having to represent the potentially infinite number of noneffects of actions. When an AI device acts, its environment changes in some ways but remains constant in countless other ways. How can an AI device update its database to reflect these changes? "If one simply excludes the non-effects from the program and only represents the effects of the actions then the problem is that it is not a matter of logic that everything else does in fact stay the same. Though it may be a matter of common sense, the artificial system cannot make this deductive inference from the limited information it possesses. If one then opts to include the non-effects in the program, the problem is that this quickly becomes computationally intractable, because the number of non-effects one must include is staggeringly large and leads to a combinatorial explosion." Madeleine Ransom, "Why Emotions Do Not Solve the Frame Problem," in *Fundamental Issues of Artificial Intelligence*, ed. Vincent C. Müller (Cham, Switzerland: Springer, 2016), 355.

34   Boden, *AI*, 154–55.

35   Ibid., 43–44. See also Ransom, "Why Emotions Do Not Solve the Frame Problem," 355.

36   Boden, *AI*, 154.

37   Thomas Dietterich's comments, in Guia Maire Del Prado, "Experts Explain the Biggest Obstacles to Creating Human-Like Robots," *Business Insider: Tech News*, March 9, 2016, https://www.businessinsider.com.au/experts-explain-the-biggest-obstacles-to-creating-human-like-robots-2016-3.

38   Erik J. Larson, "Transhumanist Claims Aside, Enhancing Human Intelligence Isn't on the Horizon," *Evolution News*, April 27, 2015, https://evolutionnews.org/2015/04/transhumanist_c.

39   Michael N. Keas, "Systematizing the Theoretical Virtues," *Synthese* 195, no. 6 (2018): 2761–93. Open access: dx.doi.org/10.1007/s11229–017–1355–6.

40   Larson, "Transhumanist Claims Aside, Enhancing Human Intelligence Isn't on the Horizon."

41   Alessio Plebe and Pietro Perconti, "The Slowdown Hypothesis," in *Singularity Hypotheses: A Scientific and Philosophical Assessment*, ed. Amnon H. Eden (Dordrecht: Springer, 2012), 356.

42   Walsh, "The Singularity May Never Be Near," 3.

43   Ibid.

44   Ibid.

45   Ibid.

46   Andrew Majot and Roman Yampolskiy, "Diminishing Returns and Recursive Self Improving Artificial Intelligence," in *The Technological Singularity: Managing the Journey*, ed. Victor Callaghan et al. (Berlin: Springer, 2017), 143.

47   Plebe and Perconti, "The Slowdown Hypothesis," 360.

48   Ibid.

49   Robert M. Geraci, *Apocalyptic AI: Visions of Heaven in Robotics, Artificial Intelligence, and Virtual Reality* (Oxford: Oxford University Press, 2010), 22 and 54.

50   Boden, *AI*, 153.

51   "It's science fiction. I don't see any particular reason to believe it." Noam Chomsky in an interview, when asked about Kurzweil's Singularity writings, 19:15–19:23, "Noam Chomsky: The Singularity is Science Fiction!," *Singularity Weblog*, October 4, 2013, https://www.youtube.com/watch?v=0kICLG4Zg8s.

52   John Horgan, "The Consciousness Conundrum," *IEEE Spectrum*, June 1, 2008, https://spectrum.ieee.org/biomedical/imaging/the-consciousness-conundrum.

53   "It is likely...that there are no such other civilizations.... That's right, our humble civilization with its pickup trucks, fast food, and persistent conflicts (and computation!) is in the lead in terms of the creation of complexity and order in the universe." Kurzweil, *The Singularity Is Near*, 357.

54   Ibid., 364.

55 Ibid., 348–62. For a more extensive review of excuses for why ET has not phoned us yet, see Webb, *If the Universe Is Teeming with Aliens.*

56 Kurzweil, *The Singularity Is Near,* 342.

57 Ibid., 485.

58 Paul Davies, "E.T. And God—Could Earthly Religions Survive the Discovery of Life Elsewhere in the Universe?" *Atlantic Monthly,* September 2003, 115.

59 C. S. Lewis, "Religion and Rocketry," in *The World's Last Night, and Other Essays* (New York: Harcourt Brace and Company, 2017), 83–92. This essay is a reprint of "Will We Lose God in Outer Space," *Christian Herald,* April 1958.

Chapter 8: Creating ET: Science Fiction as Futuristic Myth

1 Kepler's lunar story began (historically) in 1593 when he was a student at Tübingen University intending to complete a terminal degree in theology. He wrote a disputation on what would it be like to view the cosmos from the moon. Kepler would later express his pro-Copernican thought experiment in his fictional dream. While he composed most of the story in 1609, Kepler worked on the rest of his life, adding copious scientific and authorial-intent footnotes to explain many of the story's components. Kepler died unexpectedly in 1630 soon after the beginning stages of typesetting the finished manuscript, which he had entitled *Somnium seu Astronomia Lunari* (Dream or Astronomy of the Moon). Kepler's son Ludwig finally managed to publish the book in 1634.

2 Kepler's footnote after "Daemon" reads: "Knowledge of the phenomena of the heavenly bodies; from *daiein,* meaning 'to know.'" The premier English translation of Kepler's story, Edward Rosen, *Somnium; the Dream, or Posthumous Work on Lunar Astronomy* (Madison: University of Wisconsin Press, 1967), 62, explains that Kepler's derivation of Daemon from *daiein* is not accepted by modern philology. In another footnote by Kepler we learn that the lunar intelligent beings (Daemons) are allegorical for "the sciences in which the causes of phenomena are disclosed" (as quoted in Rosen, *Somnium,* 50).

3 Frederique Ait-Touati, *Fictions of the Cosmos: Science and Literature in the Seventeenth Century,* trans. Susan Emanuel (Chicago; London: University of Chicago Press, 2011), 20–23.

4 Rosen, *Somnium,* 15. Here, and in quotations of Rosen's translation of Kepler below, I have removed all footnote numbers inserted by Kepler. I refer to these footnotes only when needed.

5 Ibid., 16.

6 Ibid.

7 Dean Swinford, "The Lunar Setting of Johannes Kepler's *Somnium,* Science Fiction's Missing Link," in *Classical Traditions in Science Fiction,* ed. Brett M. Rogers and Benjamin Eldon Stevens (Oxford: Oxford University Press, 2015), 28 and 34.

8 Rosen, *Somnium,* 27.

9 Ibid., 129–32.

10 Ibid., 21.

11 Ibid., 22.

12 As viewed from the middle of the Volva-facing side of Levania, their Volva is directly overhead (zenith). If they travel in any direction from that point, then their Volva would appear progressively closer to the horizon, similar to how earthlings use the altitude of the North Star (end of the handle of the Little Dipper) to determine one's latitude.

13 Rosen, *Somnium,* 22.

14 Ibid., 36.

15 H. G. Wells, *The War of the Worlds* (London: Heinemann, 1898). This first edition of the book is available at https://archive.org/stream/warofworlds00welluoft#page/240/

mode/2up. Danielson, *Book of the Cosmos*, 178, identified Robert Burton's 1621 *Anatomy of Melancholy* as the "monumentally digressive" cosmological musings of a "learned non-scientist." One should not expect to find Kepler in holistic accuracy in such a book.

16 Wells, *The War of the Worlds*, 240–41.
17 Robert Crossley, "H. G. Wells, Visionary Telescopes, and the 'Matter of Mars,'" *Philological Quarterly* 83, no. 1 (2004).
18 Michael J. Crowe, *The Extraterrestrial Life Debate, 1750–1900* (New York: Dover, 1999), 494.
19 H. G. Wells, "Intelligence on Mars," *Saturday Review*, as cited in David C. Smith, *H. G. Wells: Desperately Mortal: A Biography* (New Haven: Yale University Press, 1986), 64–65.
20 H. G. Wells interview, *Weekly Sun Literary Supplement*, December 1, 1895, as cited in Steven McLean, *The Early Fiction of H. G. Wells: Fantasies of Science* (New York: Palgrave Macmillan, 2009), 1–2.
21 H. G. Wells, *The War of the Worlds* (London: Heinemann, 1898), 1.
22 Ibid., 2.
23 Crowe, *The Extraterrestrial Life Debate, 1750–1900*.
24 Wells, *The War of the Worlds*, bk. 2, chap. 2.
25 Ibid., bk. 1, chap. 13.
26 Michael Sherborne, *H. G. Wells: Another Kind of Life* (London: Peter Owen, 2010), 105.
27 Ibid., 170.
28 Ibid., 332–33.
29 Ibid., 333.
30 C. S. Lewis, *That Hideous Strength: A Modern Fairy-Tale for Grown-Ups* (New York: Collier, 1965), 43–44.
31 H. G. Wells, *The Shape of Things to Come* (New York: The Macmillan Company, 1933), bk. 4, chap. 1.
32 Wells published the movie script a year before the movie's debut: H. G. Wells, *Things to Come; a Film Story Based on the Material Contained in His History of the Future "The Shape of Things to Come"* (London: Cresset Press, 1935). Wells's script for the closing scene is virtually identical to the one that made it into the movie (I have quoted the final movie version here).
33 For an analysis of such mythmaking in popular science and sci-fi, see James A. Herrick, *Scientific Mythologies: How Science and Science Fiction Forge New Religious Beliefs* (Downers Grove, IL: IVP Academic, 2008).
34 I thank John Angus Campbell for helping me to formulate this way of expressing my thesis.

Chapter 9: Preaching Anti-theism on TV: *Cosmos*

1 This cosmic parody of God was intentional. Sagan grew up in a Reform/Conservative Jewish home, knew the Bible well, and then rejected theism as a teenager.
2 Although I cited Tyson's "cosmic perspective" sermon in chapter 6 as a foreword to an astronomy textbook, he first introduced this trademark in Neil deGrasse Tyson, "Universe: The 100th Essay—the Cosmic Perspective," *Natural History* (2007): 22.
3 Keay Davidson, *Carl Sagan: A Life* (New York; Toronto: J. Wiley, 2000), 36. Sagan noted the spike in UFO sightings after the United States dropped two atomic bombs on Japan. Maybe alarmed aliens had come to investigate our dangerous technological transition, he thought. On Sagan's UFO beliefs, see Davidson, *Carl Sagan*, 45–52.
4 Ibid., 37, 45.
5 Unless otherwise indicated, all the Sagan quotations below are from his 1980 *Cosmos* TV series, which are available online in many locations. Sagan, 1980 *Cosmos* TV series, ep. 1, https://www.youtube.com/watch?v=2LeCgTcuFik.

6   This is repeated in the eighth and thirteenth episodes. Numerous musicians have sung this line. It is well entrenched in pop culture. The website http://carlsagan.com uses this sentence as its tagline.

7   John G. West, "Introduction," in *The Magician's Twin: C. S. Lewis on Science, Scientism, and Society*, ed. John G. West (Seattle: Discovery Institute Press, 2012), 20.

8   Schrempp, *The Ancient Mythology of Modern Science*, 223.

9   For similar accounts of Sagan's *Cosmos* series, see Thomas M. Lessl, "Science and the Sacred Cosmos: The Ideological Rhetoric of Carl Sagan," *Quarterly Journal of Speech* 71, no. 2 (1985): 175–87; Karen Jane Sorensen, "Carl Sagan's Cosmos: The Rhetorical Construction of Popular Science Mythology" (PhD diss., North Dakota State University of Agriculture and Applied Science, 2013).

10   Matthew 4:19. Matthew reports that Peter and his brother Andrew followed Jesus immediately, leaving the seashore for a greater horizon. Davidson, *Carl Sagan*, 56, analyzes Sagan's naively optimistic "messianic" outlook: "How could one continue to believe in Progress after Germany, once the citadel of European culture and science, had stooped so low? If Carl Sagan had confronted—truly confronted—the meaning of the Holocaust, could he have remained so boyishly optimistic, so sure of the virtues of Reason and the inevitability of Progress? Without that self assurance, his subsequent career might have been quite different—still brilliant, no doubt, but surely less messianic."

11   David Paul Rebovich, "Sagan's Metaphysical Parable," *Society* 18, no. 5 (1981): 93.

12   Sagan had a reputation for arrogance, and intense interpersonal conflict, among some of his closest friends. Based on extensive interviews, Davidson reports that Sagan's "friends and family tell plenty of stories that reveal he could be overbearing and witheringly arrogant in plenty of other, less stressful situations [than the filming of *Cosmos*]." Davidson, *Carl Sagan*, 325.

13   Walker Percy, *Lost in the Cosmos: The Last Self-Help Book* (New York: Pocket Books, 1983), 201.

14   Ibid., 173.

15   Sagan, *Cosmos* TV series, ep. 13.

16   Shank, "That There Was No Scientific Activity Between Greek Antiquity and the Scientific Revolution," in *Newton's Apple*, 8.

17   Edward J. Watts, *Hypatia: The Life and Legend of an Ancient Philosopher* (New York: Oxford University Press, 2017), chap. 3.

18   Ibid., chap. 4.

19   Ibid., chap. 8.

20   Ibid., 117.

21   She remained a virgin. For her, sex was taboo. Ibid., 74.

22   Ibid., 116.

23   Ibid., 15.

24   Ibid., 13.

25   Ibid., 15. Watts similarly notes: "The fate of the Library after the third century is equally mysterious. By the fourth century, however, the Serapeum and its library had become the biggest repository of literature in the city. Whatever dedicated bibliographic resources the Museum still possessed during Hypatia's life did not compare either to those available to its predecessor or to what the Serapeum held."

26   "The Richard Dawkins Award," Atheist Alliance of America, https://www.atheist-allianceamerica.org/the-richard-dawkins-award/.

27   Unless otherwise indicated, all of the Tyson quotations below are from his 2014 *Cosmos* TV series, which are available online in many locations, including Spanish editions. See, for example, "La historia de Giordano Bruno," https://vimeo.com/89241669.

28   Davidson, *Carl Sagan*, chap. 13.

29   Corey S. Powell, "Defending Giordano Bruno: A Response from the Co-Writer of

Cosmos," Discover Magazine Blog: Out There, March 13, 2014, http://blogs.discover-magazine.com/outthere/2014/03/13/cosmos-giordano-bruno-response-steven-soter/#. WoeMmWaZPSK.

30   "La historia de Giordano Bruno," https://vimeo.com/89241669.

31   David Klinghoffer, ed., *The Unofficial Guide to Cosmos: Fact and Fiction in Neil deGrasse Tyson's Landmark Science Series* (Seattle: Discovery Institute Press, 2014).

32   "Full Show: Neil deGrasse Tyson on the New Cosmos," *Moyers and Company*, January 10, 2014, as cited in Klinghoffer, *Unofficial Guide to "Cosmos,"* 160.

33   Lennox did so as a response to Stephen Hawking and Leonard Mlodinow's book *The Grand Design* (2010). See John C. Lennox, *God and Stephen Hawking: Whose Design Is It Anyway?* (Oxford: Lion, 2010).

34   Joseph Martin, "We Need to Talk About *Cosmos*," *H-PhysicalSciences*, May 14, 2014, https://networks.h-net.org/node/25318/discussions/26537/we-need-talk-about-cosmos.

35   But interdisciplinary college courses about science, history, philosophy, and culture might be enhanced by showing clips of *Cosmos* 1980 and 2014, provided that this is done alongside responsible scholarship. I have taught such courses for a quarter century.

36   Martin, "We Need to Talk About *Cosmos*.".

37   Brannon Braga, "*Star Trek* as Atheist Mythology," *YouTube*, June 25, 2006, http://www.youtube.com/watch?v=iJm6vCs6aBA.

38   Marshall Honorof interview of Brannon Braga, "Rebooting Cosmos," *Yahoo News*, January 14, 2014, as cited in Klinghoffer, *Unofficial Guide to "Cosmos,"* 161.

Chapter 10: Kepler, Devout Scientist

1   This occurs halfway through Sagan, 1980 *Cosmos* TV series, ep. 3.

2   Sagan's leading biographer supports my thesis of Sagan's projecting himself onto Kepler. "Sagan may have concentrated on Kepler for another, more personal reason: he may have glimpsed aspects of himself in the astronomer-astrologer. Like Kepler, Sagan was a contradictory figure, a man with one foot in science and the other in imagination, one in logic and the other in Edgar Rice Burroughs [creator of Tarzan and a prolific science-fiction author].... Sagan's handling of Kepler constitutes part of a broader habit of Sagan's that might be called 'covert autobiography.' His books include many mini-biographies of great scientists. These superb biographical essays emphasize numerous minor aspects of the subject's life, perhaps because, although Sagan never says so, they reminded him of himself.... In reality, he may have been seeking support or consolation for his own eccentricities or failings by spotting them in the lives of scientists far greater than he." Davidson, *Carl Sagan*, 335–36.

3   Davidson, *Carl Sagan*, 11–15, explores Sagan's Jewish family. Carl remembered more of the Reform (liberal) side of growing up: a father who ate bacon outside the home and was uninterested in religion. His sister, Cari, remembered more of the Conservative Jewish side of the family in their mother. Cari told Davidson during an interview that her mother "definitely believed in God and was active in the temple."

4   Max Caspar et al., eds., *Johannes Kepler Gesammelte Werke* (Munich: C. H. Beck, 1937–), *Johannes Kepler Gesammelte Werk*, vol. 18, no. 1146, lines 33–34: "confidenter respondebat, unico salvatoris nostri Jesu Christi merito." Letter from Stephan Lansius in Regensburg (where Kepler died) to an anonymous recipient in Tübingen, 24 January 1631 (old style calendar). *Merito* in this letter may be translated "merit," "service," or "good work," according to classicist Donald Kim in an April 9, 2013, email. *Merito* is the substantive participle formed from *mereo*, the verb "to deserve, merit."

5   Johannes Kepler, *Epitome of Copernican Astronomy*, as cited in Max Caspar, *Kepler*, trans. C. Doris Hellman (New York: Dover, 1993), 381.

6   Ibid.

7   Kepler's textbook dedication, as translated in Johannes Kepler and Carola Baumgardt, *Johannes Kepler: Life and Letters* (New York: Philosophical Library, 1951), 122–23.

8   Johannes Kepler, *Epitome of Copernican Astronomy*, trans. Charles Glenn Wallis, Great Books of the Western World, vol. 16 (Chicago: Encyclopedia Britannica, 1952), 849–50.

9   Emma Willard, *Astronomy and Astronomical Geography* (New York: A. S. Barnes and Burr, 1860), 254. Although Kepler did not use the scientific term *laws* to describe his discoveries, he thought he was discovering important truths about the physical world (what *we* call "natural laws"). His most notable successes were retrospectively called the "three laws of planetary motion."

10 Kepler dissented from a few Lutheran doctrines, especially certain details about the Eucharist.

11 Although a 1901 autopsy found traces of mercury in hairs from Tycho's beard, the subsequent "Kepler poisoned Tycho" rumor was undercut by a 2010–12 autopsy that showed insufficient mercury or other toxins to trigger death.

12 Kepler, 1610 published letter, in Johannes Kepler and Edward Rosen, *Kepler's Conversation with Galileo's Sidereal Messenger* (New York: Johnson Reprint, 1965), 43.

13 Owen Gingerich, "Johannes Kepler," in *Dictionary of Scientific Biography*, ed. Charles Coulston Gillispie and American Council of Learned Societies, vol. 7 (New York: Scribner, 1970), 295.

14 Bernard R. Goldstein and Hon Giora, "Kepler's Move from Orbs to Orbits: Documenting a Revolutionary Scientific Concept," *Perspectives on Science* 13, no. 1 (2005).

15 He discovered that the sun, like a fireplace located near the center of a house, is located at one of the two foci. Focus means "hearth" in Latin.

16 Kepler discovered this law before his first law. Later astronomers renumbered them to make them easier to teach.

17 Kepler to J. G. Herwart von Hohenburg, April 9/10, 1599, as translated in Baumgardt, *Kepler: Life and Letters*, 50.

18 Kepler to J. G. Herwart von Hohenburg, February 16, 1605, in Caspar, *Johannes Kepler Gesammelte Werke*, vol. 15, no. 325, lines 57–61.

19 Patrick J. Boner, *Kepler's Cosmological Synthesis: Astrology, Mechanism, and the Soul* (Leiden: Brill, 2013).

20 Johannes Kepler, *The Harmony of the World*, trans. A. M. Duncan, E. J. Aiton, J. V. Field (Philadelphia: American Philosophical Society, 1997), 115. In saying God created the world "in weight, measure, and number," Kepler alludes to the apocryphal *Book of Wisdom*, 11:20 (contained in the 1611 King James Bible), in which God is said to have "ordered all things in measure and number and weight."

21 Christopher B. Kaiser, *Toward a Theology of Scientific Endeavour: The Descent of Science*, Ashgate Science and Religion Series (Aldershot, England: Ashgate, 2007), 51, 143.

22 Ibid., 48.

23 William A. Dembski et al., *The Patristic Understanding of Creation: An Anthology of Writings from the Church Fathers on Creation and Design* (Riesel, TX: Erasmus Press, 2008), 91–92.

24 Ibid., 27.

25 John 1:3, ESV.

26 Nicholas Jardine and Johannes Kepler, *The Birth of History and Philosophy of Science: Kepler's "A Defence of Tycho Against Ursus"* (Cambridge: Cambridge University Press, 1984), 139.

27 Kepler's 1600 treatise was unpublished and unknown until the manuscript was discovered in the nineteenth century. But in a work published in 1596, he articulated some of his thoughts about what are now called the theoretical virtues: Johannes Kepler, *Mysterium Cosmographicum: The Secret of the Universe*, trans. A. M. Duncan (New York: Abaris, 1981). Unlike previous defenders of the Copernican system, in this book Kepler asserted not only its mathematical possibility but also its physical reality.

28 Kepler pioneered the study of theoretical virtues and its larger disciplinary framework, the history and philosophy of science. He did this by philosophical reflection on his own scientific work and that of his predecessors. Kepler also relied on certain theological premises as he constructed some of the foundations for this field of study. I have built on Kepler's foundational work in recognizing the virtues of a good theory by systematizing these theoretical virtues as they have emerged even more clearly in the practice of science since Kepler. Michael N. Keas, "Systematizing the Theoretical Virtues," *Synthese* 195, no. 6 (2018): 2761–93. Open access: dx.doi.org/10.1007/s11229-017-1355-6. I identify and classify twelve traits that are critical to assessing whether a theory is plausible or even beyond reasonable doubt. Following Kepler's pioneering work of 1600, I use pivotal historical case studies in my argument. Many of the theoretical virtues operate legitimately in evaluating theological theories as well as theories that aim to explain a sequence of events in human history. But my focus in this essay is on theory evaluation in the natural sciences.

29 Jardine and Kepler, *The Birth of History and Philosophy of Science*, 140.

30 Ibid., 141–42.

31 Although others before Kepler gave some attention to the causal adequacy of theories of the heavens, Kepler did so more rigorously. Kepler's scientific achievement was supported by his own philosophical reflection on the theoretical virtues.

32 Kepler, *Mysterium Cosmographicum*, 55.

33 Rosen, *Kepler's Conversation with Galileo's Sidereal Messenger*, 43.

34 Ibid., 39.

35 Robert S. Westman, *The Copernican Question: Prognostication, Skepticism, and Celestial Order* (Berkeley: University of California Press, 2011), 400.

36 Kepler did all this when he discovered three laws of planetary motion in response to the planetary data he had inherited from Tycho. Of course, Kepler, like most scientists, speculated about many other things that turned out to be wrong.

37 Westman, *The Copernican Question*, 400.

38 Ibid.

39 Ibid.

40 Rosen, *Kepler's Conversation with Galileo's Sidereal Messenger*, 39.

41 Ibid., 37.

42 Ibid., 34.

43 Kepler to Johann Georg Brengger, November 30, 1607, as translated in Baumgardt, *Kepler: Life and Letters*, 77–78.

44 Michael J. Crowe, *The Extraterrestrial Life Debate, 1750–1900* (New York: Dover, 1999), 11.

45 Rosen, *Kepler's Conversation with Galileo's Sidereal Messenger*, 11.

46 Kepler to Johann Georg Brengger, April 5, 1608, as translated in Baumgardt, *Kepler: Life and Letters*, 78–79.

47 Max Caspar, *Kepler*, trans. C. Doris Hellman (New York: Dover, 1993), 385.

48 Historian of science Edward Davis in personal email of September 14, 2018, used by permission, prefers to "avoid the term 'revisionist'" for "those who have shaped the Bruno story (or any other story) to attack Christianity." His advice stems from the unthoughtful use of "revisionist" by certain public voices who "know only enough history to be dangerously misleading." Appropriately, Davis notes that "the term is ideologically neutral. It doesn't equate with the bad guys in one's narrative. Using it exclusively as a pejorative implies that there was originally a true history that is later rewritten by purveyors of false facts and must now be recovered. That begs questions about the authors of that original, 'true history'—about their motives, their biases, and also their state of knowledge, which is not necessarily superior to the historians who come along later. Perhaps on some matters they did know more than us—thus we regard them as primary sources—but, on other matters maybe they didn't, and maybe genuine new facts will cause us to rewrite

their accounts." I fully accept Davis's wise cautionary note about using the term *revisionist*. Accordingly, it occurs but once in my book, here in this Kepler chapter, and for good reasons. Kepler is a reliable primary source about Bruno. Compare chapters 4 and 10 to see why I reached this conclusion. Moreover, Kepler helped pioneer the very field that Davis and I both practice: the history of science.

Chapter 11: Remembering America's Harmony of Science and Faith

1   Samuel Eliot Morison, "The Harvard School of Astronomy in the Seventeenth Century," *New England Quarterly* 7 (1934): 6. Some Harvard students had access to this later (posthumous) edition of *Physiologia peripatetica*.
2   Library Harvard College, W. H. Bond, and Hugh Amory, *The Printed Catalogues of the Harvard College Library, 1723–1790* (Boston: Oak Knoll Press, 1996), 89. The library, founded in 1638, acquired all three volumes of Kepler's textbook.
3   Vincent Wing, *Astronomia Instaurata, or, a New Compendious Restauration of Astronomie in Four Parts* (London: Company of Stationers, 1656).
4   Morison, "The Harvard School of Astronomy," 9. Morison found that the copy of Wing's 1656 textbook in the Harvard library "bears marks of ownership" of the following people listed by year of graduation: Thomas Graves (AB, 1656), Samuel Brackenbury (AB, 1664), Joseph Browne (AB, 1666), and Edward Holyoke (AB, 1705). Browne wrote the New England almanacs published at the Harvard press for the years 1667 and 1669, continuing a tradition of Harvard tutors serving the New England community.
5   Ibid., 11. Morison provides a complete transcription of Brigden's Copernican almanac essay.
6   Ibid., 9.
7   Donald K. Yeomans, "The Origin of North American Astronomy—17th Century," *Isis* 68 (1977): 414–15. Yeomans maintains that almanacs "remained the only American periodical literature in the 17th century," and that "together with Puritan religious sermons, almanacs made up the majority of all 17th-century American literature."
8   Morison, "The Harvard School of Astronomy," 10.
9   "Samuel Eliot Morison," in *Naval History and Heritage Command*, undated biography, https://www.history.navy.mil/content/history/nhhc/research/histories/biographies-list/bios-m/morison-samuel-e.html.
10 Morison, "The Harvard School of Astronomy," 4. He also perpetuated the myth that premodern astronomers in the Western tradition believed in a tiny cosmos.
11 Shortly after publishing the 1659 almanac, Brigden served as a preacher in Stonington, Connecticut, until his death at age twenty-three. John Langdon Sibley et al, *Biographical Sketches of Graduates of Harvard University*, 3 vols., vol. 1 (Cambridge, MA: Charles William Sever, University Bookstore, 1873), 496.
12 I studied all the English-language astronomy textbooks that I could find in all the Harvard catalogues up to 1790. This included all such textbooks listed under the heading of "astronomia" in the 1790 subject-organized catalog, and such textbooks as I could locate from the two earlier catalogues (1723 and 1773) that were organized alphabetically by author. The 1723 printed catalogue is the only one prior to the 1764 fire that destroyed most of the Harvard Library. Although Wing's *Astronomia instaurata* is not listed in any of these printed library catalogues up to 1790, I noted earlier in this essay how we know that several early Harvard students owned and used Wing's astronomical reference book.
13 Johannes Kepler, *New Astronomy*, trans. William H. Donahue (Cambridge: Cambridge University Press, 1992), 59. In the footnote here, Donahue writes: "The following arguments on the interpretation of scripture were to become the most widely read of Kepler's

writings. They were often reprinted from the seventeenth century on, and translated into modern languages. Indeed, this part of the Introduction was the only work of Kepler's to appear in English before 1700."

14 Thomas Salusbury and Stillman Drake, eds., *Mathematical Collections and Translations*, vol. 1 (London: Dawsons, 1967). It includes translations of Galileo's 1615 letter to the Grand Duchess Christina and parts of Kepler's *Astronomia nova* (1609).

15 Joseph Moxon, *A Tutor to Astronomy and Geography. Or, the Use of the Copernican Spheres in Two Books* (London: Printed for the author, 1665), 12.

16 Ibid., 12–13.

17 Johannes Kepler, *Selections from Kepler's "Astronomia Nova,"* trans. William H. Donahue, Science Classics Module for Humanities Studies (Santa Fe: Green Cat Books, 2004), 22.

18 Ibid., 23.

19 Ibid., 24.

20 Moxon, *A Tutor to Astronomy*, 14.

21 Kepler, *Selections from Kepler's Astronomia Nova*, 20.

22 Ibid., 19.

23 David Gregory, *The Elements of Astronomy: Physical and Geometrical*, 2 vols. (London: John Morphew, 1715), i.

24 Ibid., xiv, 132–37, 145–50, 200, 376, 811, and 847.

25 Ibid., 134–35.

26 William Whiston, *Astronomical Lectures, Read in the Publick Schools at Cambridge* (London: R. Senex, 1715), 38.

27 Ibid., 13–17.

28 John Keill, *An Introduction to the True Astronomy: Or, Astronomical Lectures Read in the Astronomical School of the University of Oxford* (London: Henry Lintot, 1721), 16–17.

29 Isaac Watts, *The Knowledge of the Heavens and the Earth Made Easy: Or, the First Principles of Astronomy and Geography* (London: J. Clark and R. Hett, 1726), v. I cite the "corrected" second edition of 1728.

30 Ibid., vi.

31 Ibid., vi–vii.

32 Roger Long, *Astronomy, in Five Books* (Cambridge: Printed for the author, 1742), vii. Long's second installment of this series of books was issued in 1764, and the work was completed posthumously in 1784 by Richard Dunthorne.

33 Ibid., 725.

34 Richard Baum, "Ferguson, James," in *The Biographical Encyclopedia of Astronomers*, ed. Thomas A. Hockey et al. (New York: Springer, 2014).

35 "American publishers offered more editions of Ferguson's astronomy textbooks in the early republic than those of any other author.... Authors of other textbooks recommended that readers interested in a more extensive education in the subject consult 'Ferguson's Astronomy.' And colleges such as the University of North Carolina and Princeton regarded his textbooks as required reading." Lily A. Santoro, "The Science of God's Creation: Popular Science and Christianity in the Early Republic" (PhD diss., University of Delaware, 2011), 78–79.

36 James Ferguson, *Astronomy Explained upon Sir Isaac Newton's Principles* (London: Printed for the author, 1756), 58.

37 Ibid., 194–95.

38 Ibid., 3, 6, 26.

39 William Emerson, *A System of Astronomy: Containing the Investigation and Demonstrations of the Elements of That Science* (London: J. Nourse, 1769), viii.

40 Ibid., viii–ix.

Chapter 12: Telling the Larger Truth

1   For example, Don Page cites Psalm 8 for its implicit recognition "that a human is far smaller than the heavens," and yet the Psalmist responded with praise to God for "human glory." Don N. Page, "Our Place in the Vast Universe," in *Science and Religion in Dialogue*, ed. Melville Y. Stewart (Malden, MA: Wiley-Blackwell, 2010). Page, as distinguished university professor at the University of Alberta, works in quantum cosmology and theoretical gravitational physics. He was a doctoral student of, and published several journal articles with, the eminent Stephen Hawking.

2   Psalm 8:1b–5, ESV.

3   Michael J. Crowe, "Astronomy and Religion (1780–1915): Four Case Studies Involving Ideas of Extraterrestrial Life," *Osiris* 16 (2001): 209–26. Crowe shows how the Christian convictions of the English astronomer Edward Walter Maunder of the Royal Observatory Greenwich preconditioned him to be more critical of the claims of canals on Mars built by Martians. He was a leading figure in recognizing new scientific evidence against the ET-technology interpretation of canal-like features on Mars. In the conclusion of one of his scientific studies of Mars, he quoted part of Psalm 8. See also Crowe, *The Extraterrestrial Life Debate, 1750–1900*, for a larger study of other astronomers and theologians appealing to Psalm 8 and 19 (and similar biblical passages) in the ways that I have outlined.

4   Here are the only current astronomy textbooks in my sample that avoid all the myths surveyed in my book: Stacy E. Palen et al., *Understanding Our Universe* (New York: W. W. Norton, 2012); Laura Kay, *21st Century Astronomy*, 4th ed. (New York: W. W. Norton, 2013); Neil F. Comins and William J. Kaufmann, *Discovering the Universe*, 10th ed. (New York: W. H. Freeman, 2014).

5   Dennis R. Danielson, "The Bones of Copernicus," *American Scientist* 97, no. 1 (2009): 50–57.

6   Danielson, "Copernican Cliché," 1033. There is insufficient biographical information about these astronomers who perpetuate the Copernican devotion myth. So it is difficult to document what Danielson suspects here.

7   Franklyn Mansfield Branley et al., *Astronomy* (New York: Crowell, 1975), 517. This textbook never made it past the first edition. Today Branley is primarily remembered as the author of more than a hundred science books for children, many of which are still in print. This is where his talent as a writer and science educator still shines. Kenneth L. Franklin, obituary of Franklyn M. Branley (1915–2002), American Astronomical Society, 1455–57, http://articles.adsabs.harvard.edu/cgi-bin/nph-iarticle_query?2003BAA S...35.1455F&data_type=PDF_HIGH&whole_paper=YES&type=PRI NTER&filetype=.pdf.

8   Richard Dawkins, *The God Delusion* (Boston: Houghton Mifflin Co., 2006), 72.

9   Michael Shermer, *Skeptic: Viewing the World with a Rational Eye* (New York: Henry Holt, 2016), 125. See also "Shermer's Last Law," January 2002, http://www.michaelshermer.com/2002/01/shermers-last-law/.

10 Ray Kurzweil, *The Singularity Is Near: When Humans Transcend Biology* (New York: Viking Books, 2005), 5.

11 The only exception to this rule is when I quote a textbook author's website (http://www.adamfrankscience.com), where he identifies himself as an "evangelist of science."

12 Jeffrey Bennett et al., *The Cosmic Perspective*, 8th ed. (Boston: Pearson, 2017), 3.

13 Michael A. Seeds and Dana E. Backman, *Foundations of Astronomy*, 13th ed. (Boston: Cengage Learning, 2017), 628.

14 Eric Chaisson and Steve McMillan, *Astronomy Today*, 9th ed. (Boston: Pearson, 2018), 726. This is part of a small-group-discussion activity for students at the end of the chapter.

15 Adam Frank et al., *Astronomy: At Play in the Cosmos* (New York: W. W. Norton, 2016), 5–6.

16 http://www.adamfrankscience.com.
17 Frank, *Astronomy*, 246.
18 Timothy F. Slater and Roger A. Freedman, *Investigating Astronomy: A Conceptual View of the Universe*, 2nd ed. (New York: W. H. Freeman, 2014), 201.
19 Jay M. Pasachoff and Alexei V. Filippenko, *The Cosmos: Astronomy in the New Millennium*, 4th ed. (Cambridge: Cambridge University Press, 2013), 553.
20 Ibid., 555.
21 Sandra M. Faber, "Why Astronomy?" in Roger A. Freedman et al., *Universe*, 10th ed. (New York: W. H. Freeman, 2014), 20.
22 Kari Paul, "Merry Christmas! Why Millennials Are Ditching Religion for Witchcraft and Astrology," *MarketWatch*, December 19, 2017, https://www.marketwatch.com/story/why-millennials-are-ditching-religion-for-witchcraft-and-astrology-2017-10-20. "Meanwhile, more than half of young adults in the U.S. believe astrology is a science compared to less than 8% of the Chinese public. The psychic services industry—which includes astrology, aura reading, mediumship, tarot-card reading and palmistry, among other metaphysical services—grew 2% between 2011 and 2016. It is now worth $2 billion annually, according to industry analysis firm IBIS World."
23 "Religious Landscape Study," Pew Research Center, 2015, Chart: Age Distribution, http://www.pewforum.org/religious-landscape-study/age-distribution/.
24 Katie J. M. Baker, "Hexing and Texting," *Newsweek*, October 24, 2013, http://www.newsweek.com/2013/10/25/hexing-texting-243730.html.
25 Sagan, *Cosmos* 1980 TV series, ep. 1, https://www.youtube.com/watch?v=2LeCgTcuFik.
26 Joseph O. Baker and Christopher D. Bader, "A Social Anthropology of Ghosts in Twenty-First-Century America," *Social Compass* 61, no. 4 (2014): 571.
27 C. S. Lewis foresaw this in his final space novel, *That Hideous Strength* (1945).
28 Davidson, *Carl Sagan*, 36.
29 Dawkins, *The God Delusion*, 72.
30 Edward B. Davis, "Christianity and Early Modern Science: The Foster Thesis Reconsidered," in *Evangelicals and Science in Historical Perspective*, ed. David N. Livingstone et al. (Oxford: Oxford University Press, 1999).
31 Peter Harrison, *The Fall of Man and the Foundations of Science* (Cambridge: Cambridge University Press, 2007).

Appendix A: The First Urban Myth of the Space Age

1 Urban myths, also called urban legends, are "urban" in that (1) they are of *recent* origin, since urban life has become the dominant way people live, and (2) they originate and proliferate through urban or *mass communication* means, including newspapers and social media. The six historic myths of my book predate the urban category of myth. The ET enlightenment myth, although recent, is not "urban" because it *originated among the cultural elite* rather than within pop culture (though it has proliferated in pop culture). It also bears the classic traits of "myth" in the anthropological sense (worldview-shaping imaginative archetypal narrative), which urban myths largely lack.
2 See the English version of the relevant part of the 2006 interview: "Did Yuri Gagarin Say He Didn't See God in Space?" *Orthodox Christianity and the World*, April 12, 3013, http://www.pravmir.com/did-yuri-gagarin-say-he-didnt-see-god-in-space/, which is translated from the Russian text at http://www.pravoslavie.ru/4944.html.
3 Andrew L. Jenks, *The Cosmonaut Who Couldn't Stop Smiling: The Life and Legend of Yuri Gagarin* (DeKalb, IL: NIU Press, 2012).
4 Victor J. Stenger, *The New Atheism: Taking a Stand for Science and Reason* (Amherst, NY: Prometheus Books, 2009), 59.
5 Walter G. Vincenti, *What Engineers Know and How They Know It: Analytical Studies from*

Aeronautical History (Baltimore: Johns Hopkins University Press, 1990). In this land-mark study, Vincenti explains how even engineers, as the most sophisticated technolo-gists, often acquire "know-how" knowledge without applying scientific theory ("know-ing that"). He analyzes several case studies in early aeronautical engineering to show that to a large degree this technological discipline existed independent of the physi-cal sciences. Sometimes engineers acquired aeronautical "know-how" knowledge (e.g., improvements in wing airfoils and propeller shapes) through sheer trial and error. More often they created and systematically used sophisticated mathematical models without applying the natural sciences.

6   "The Seeing Eye," in C. S. Lewis and Walter Hooper, *The Seeing Eye and Other Selected Essays from Christian Reflections* (New York: Ballantine Books, 1986), 225–30. For a fuller theological critique of the assumptions underlying this urban myth, see Thomas F. Torrance, *Space, Time, and Incarnation* (Oxford: Oxford University Press, 1969).

Appendix B: A New Argument for Pascal as a Copernican

1   The passage we are analyzing, as rendered in the 2010 digitized version at http://www. samizdat.qc.ca/arts/lit/Pascal/Pensees_1671_ancien.pdf, which was transcribed from the 1671 edition housed in the National Library of France, reads: "Qu'il considere cette éclatante lumiere, mise comme une lampe éternelle, pour éclairer l'Univers. *Que la terre luy paroisse comme un point au prix du vaste tour que cet astre décrit.* & qu'il s'étonne de ce que ce vaste tour luy mesme n'est qu'un point tres délicat, à l'égard de celuy que les astres qui roulent dans le firmament embrassent." The italicized part underlies Rawl-ings's accurate translation: "let the earth appear to him as a single speck compared with the vast orbit which *this star* describes." Note that "cet astre" means "this star," not "the sun," as Trotter mistranslates it.

2   The 1671 edition cited above reads, "mise comme une lampe éternelle, pour éclairer l'Univers." This offers another hint of Pascal's Copernican mindset: his sun descrip-tion echoes Copernicus, *De revolutionibus* 1.10: "For who, in this most beautiful temple, would *set this lamp* in another or a better place, whence to *illuminate all things at once*?" (emphasis mine).

3   Blaise Pascal, *The Thoughts, Letters, and Opuscules of Blaise Pascal*, trans. O. W. Wight (New York: Derby and Jackson, 1859), 158.

4   Michael J. Crowe and Matthew F. Dowd, "The Extraterrestrial Life Debate from Antiquity to 1900," in *Astrobiology, History, and Society: Life Beyond Earth and the Impact of Discovery*, ed. Douglas A. Vakoch (Berlin: Springer, 2013), 52.

Acknowledgments

Thanks to Planet Fitness, where, with iPad in hand, I did much of the heavy lifting that became this book. I thank Jed Donahue, ISI Books editor in chief, for his excellent editing. I am thankful for Kerry Magruder and David Klinghoffer, who also helped me improve the entire book. Edward Davis offered wise counsel on the first half of my book. I also acknowledge the assistance of those who read smaller portions of the book: John Bloom, John Angus Campbell, Michael Crowe, Guillermo Gonzalez, James Hannam, Timothy Heil, James Herrick, Caise Jones, David Keas, Gilbert Keas, Brian Krouse, Steve Laufmann, George Montañez, Jay Richards, Jeffrey Burton Russell, William Shea, Jacob Theiss, John West, and Jonathan Witt.

In 2012 the University of Oklahoma awarded me a Mellon travel grant to use the History of Science Collections for my project. My book largely began with this grant. I finished the project with support from Discovery Institute during the 2017–18 academic year.

Chapter 6 owes much to the editorial assistance I received when preparing "Myth #3: That the Copernican Revolution Demoted the Status of Earth," in *Newton's Apple and Other Myths About Science*, edited by Ronald Numbers and Kostas Kampourakis (Harvard University Press, 2015). I also thank my other fellow historians of science who interacted with me during the conference aimed at preparing the myth-busting essays for *Newton's Apple*.

Index

INTERCOLLEGIATE
STUDIES INSTITUTE

think. live free.™

ISI Books is the publishing imprint of the **Intercollegiate Studies Institute**.

Most thoughtful college students are sick of getting a shallow education in which too many viewpoints are shut out and rigorous discussion is shut down.

We teach them the principles of liberty and plug them into a vibrant intellectual community so that they get the collegiate experience they hunger for.

www.isi.org